Alice's Adventures in Molecular Biology

Alice's Adventures in Molecular Biology

Arieh Ben-Naim
Roberta Ben-Naim

The Hebrew University of Jerusalem, Israel

 **World Scientific**

NEW JERSEY · LONDON · SINGAPORE · BEIJING · SHANGHAI · HONG KONG · TAIPEI · CHENNAI

Published by

World Scientific Publishing Co. Pte. Ltd.

5 Toh Tuck Link, Singapore 596224

USA office: 27 Warren Street, Suite 401-402, Hackensack, NJ 07601

UK office: 57 Shelton Street, Covent Garden, London WC2H 9HE

Library of Congress Cataloging-in-Publication Data
Ben-Naim, Arieh, 1934–
   Alice's adventures in molecular biology / Arieh Ben-Naim, Roberta Ben-Naim, Hebrew University of Jerusalem, Israel.
   pages cm
   Includes index.
   ISBN 978-9814417242 (hardcover : alk. paper) -- ISBN 978-9814417259 (pbk. : alk. paper)
   1. Molecular biology--Popular works. 2. Molecular biology--Humor. I. Ben-Naim, Roberta. II. Title.
   QH506.B46 2013
   572.8--dc23
                                        2013011740

British Library Cataloguing-in-Publication Data
A catalogue record for this book is available from the British Library.

Typeset by Stallion Press
Email: enquiries@stallionpress.com

Printed by FuIsland Offset Printing (S) Pte Ltd Singapore.

Dedication

This book is dedicated to Avital, Yariv, Yair and Mirren Jean,
the future generation, and like many others their age
who we had in mind when we wrote this book.

Contents

Preface

This book is a sequel to *Alice's Adventures in Water-Land* by the same authors. Like the previous volume, this new book has two aims; first, to provide the reader with a glimpse of the relatively new field of molecular biology, and second, to demonstrate the indispensable role of water molecules in some fundamental biochemical processes.

We have melded scientific fact with fiction, injecting humor and prose, which will hopefully make the reading experience more enjoyable, yet informative, while addressing a wider range of readers. We hope that this book will arouse the curiosity of lay readers and encourage them to explore further the fascinating and rapidly developing field of molecular biology.

Arieh Ben-Naim
Roberta Ben-Naim
May 2012

Acknowledgments

We are grateful to our friends Frank Bierbrauer, Xiao Liu, Robert Mazo, Mihaly Mezei, Jorge Numata, Michael Gross, Paul King, Gerald Pollack and Raffaele Vena for reading part or the whole manuscript and offering valuable comments.

1

Alice Returns to Class

When Alice woke up on the first day of class, she literally had to drag herself out of bed. She wished she was still on vacation, but on the other hand, the idea of going back to school thrilled her. As soon as she entered the campus, allowing the familiar surroundings to slowly work their magic, her spirits soared.

From the day she had attended Professor Holmes' last lecture, and visited his laboratory, she had looked forward to coming back. She had diligently brushed up on her biology while on vacation, and was impressed by how many advances had been made. She was going to learn about the mysterious microscopic world within the realm of molecular biology, a realm quite different from macroscopic biology. This echoed a very different facet of *biology*. Studying living creatures and their behavior is just one aspect of what biology is all about. But underlying all these phenomena, a huge number of inconspicuous activities are taking place. The realm of atoms and molecules is not readily seen, but the combined effects of their activities create the phenomena that we witness with our naked eyes.

The relationship between the microscopic and the macroscopic worlds tickled Alice's imagination. She vividly recalled her experiences in the lab, and she knew how these worlds were so different, yet were only two sides, two facets, of the same phenomena.

While on vacation, she had been actively engaged in visualizing how things looked from the 'inside,' or from the microscopic point of view. In particular, she was looking forward to learning more about what was going on inside her own cells — something the professor had said was going to be part of the new term's lessons.

Aside from her eagerness to attend Professor Holmes' lessons and study molecular biology, Alice also greatly missed her forays into the 'inaccessible realms,' which only the professor's laboratory could have provided her. So vivid were her recollections of what she had seen and experienced in the 'shrinking machine' that it seemed to her that they had happened only yesterday. Although she now knew that the machine was not what she had originally been led to believe, she still found the illusion fascinating — and she was really looking forward to more adventures.

The stillness of the empty classroom didn't last long when Alice's classmates started trickling in. Girls giggled and shrieked as they talked about their summer escapades, video games whirred and bleeped, someone dribbled a ball around the classroom. But the deafening cacophony died down instantly when Professor Holmes entered from the back door. He greeted the class in his familiar booming voice: "Welcome back to my class!"

Fig. 1 Kofeau sitting in class.

But who — or rather what — followed the professor as he entered the classroom was unexpected. Some students could not control their laughter. A few others had raised eyebrows. Alice, like most of the class, was wide-eyed with amusement. A little monkey with big round eyes, dressed in a school uniform, with a mailman's bag slung diagonally across his hairy chest and a university ID loosely dangling around his neck, silently trailed the professor. Soon, the whole class was in stitches.

"Good morning! I see some familiar faces — Bob, Linda, and Alice, of course," said the professor with a knowing smile.

"Welcome to my molecular biology course. This semester we shall be learning about some very fundamental processes that go on in our bodies, or rather in the cells in each one of us — as well as in the cells of our friend here, Kofeau."

Upon hearing his name, the little monkey stood up and bowed to the class, and then sat down again without making a sound. Alice, thinking that bringing Kofeau to class had to be one of the professor's tricks, immediately started trying to figure out what he was up to.

"You must be surprised to see Kofeau in class, but there is really nothing unusual about it. This is all about biology, and our friend here is part of biology. You must have heard people say that a dog is a man's best friend, and that is perfectly all right because they are indeed our good and loyal friends. However, in some fundamental ways monkeys are much closer to us than dogs, not only in their appearance and in their behavior, but also in their 'book of life.' This book of life, stored in every cell of the human body, is more than 96% identical to that stored in Kofeau's cells. That is the reason I brought him to class, aside from him carrying my stuff," the professor said, grinning from ear to ear. "Kofeau is really the closest species to mankind."

Looking at the monkey, the professor said, "Oh, I forgot to tell you that I named him 'Kofeau' because it's a combination of 'kof' and 'eau'; *kof* is 'monkey' in Hebrew and *eau* is 'water' in French — together *Kofeau*, pronounced 'Kofo.'"

When he heard his name mentioned again, the little monkey stood to attention and looked back at Professor Holmes. He seemed to be waiting for

Fig. 2   Kofeau water content.

instructions from the professor, but when nothing came, he sank back into his chair, and nonchalantly continued his earlier pre-occupation — rolling his eyes and searching every nook and cranny of the room.

Professor Holmes continued, "For the time being, I will not explain further why I chose this particular name for our friend. For now, bear in mind that our bodies, as well as Kofeau's, contain about 70% water. But that is not all there is to the name. We shall also see how Kofeau can help us understand the role of water in some fundamental processes that occur in our cells, as well as in Kofeau's."

Professor Holmes paused briefly and threw a quick glance at Kofeau, who was sitting comfortably and quietly in the corner. However, upon seeing the students reach for their notepads, he lifted the flap of his bag, dug deep into it, and produced his own notepad — so large that it dwarfed his tiny face. Fishing inside the bag again, he produced a big, fat pencil, which was longer than his own hands. Not done yet, he opened the bag, almost tearing it apart, looking intensely for something, and with a sigh of relief found what he was looking for — oversized eyeglasses, which he quickly put on. His hilarious antics brought the house down. Even the professor couldn't suppress a chuckle. When the laughter in the room died down, he continued.

"Don't confuse what I just called the 'book of life,'" he said, "with a book that narrates the story of one's life, which we call an autobiography. The latter is a real book, as you know." When he heard the word 'autobiography,' Kofeau frantically searched inside the bag and produced yet another huge book bearing his own name on the cover, and with the book's boldly written title: *Autobiography*. He held the book up for everyone to see.

Fig. 3 Kofeau with biography.

Fig. 4 Kofeau with string of letters.

Without turning to Kofeau, the professor went on: "The 'book of life' is very different from the book your little classmate is showing you. The 'book of life' is like a long — very long in fact — sequence of letters that we cannot read. It does not tell the story of any specific person's life." Setting the autobiography aside, Kofeau's scrawny fingers plunged into the bag yet again, fishing out an immensely long ribbon. It was so long that it seemed as if would take him forever to find the other end. The class applauded thunderously, and the little monkey seemed pleased with his audience's reaction. Kofeau's ribbon revealed a sequence of letters; Alice wondered what they stood for.

"As you can see," said the professor, "what I referred to as the 'book of life' is not a book at all, and it tells no story. In fact, what you are looking at is merely a metaphor for what I referred to as the 'book of life.' This book does not tell you anything about life. It only contains information relevant to the construction of everyone's bodies, including Kofeau's. OK, show time's over." Kofeau patiently gathered both ends of the ribbon and tucked it inside the bag, and then sat quietly again.

"Let us now discuss what is meant by the 'book of life.' Of course, there is no such book stored in our cells, but there is some kind of *information* that is stored in a very long molecule called DNA. This molecule is very long in molecular terms, and it's tightly packed in the nucleus of our cells. We shall learn about the meaning of this information, the language in which this information is written, and how it is translated into instructions to make our entire bodies. As you can see, many of our features are similar to those of the primates. The similarity stems from the similarity between the DNA of our species and those of the primates. There is another branch of biology that studies the evolution of the various species. This theory will not be discussed in this course. We shall focus on the central dogma of molecular biology."

Without wasting time, Professor Holmes flashed a slide on the screen, the very same one that he showed on the first day of class last semester:

THE CENTRAL DOGMA OF MOLECULAR BIOLOGY

"This semester we shall learn about the central dogma of molecular biology and perhaps a little beyond that. There are essentially three molecules involved in the central dogma: the DNA, the RNA, and proteins. At this stage, we do not need to know the exact molecular structure of any of these molecules; remembering the initials will be sufficient for now. Briefly, the DNA is a molecule that carries all the information that is essential in building up the entire cell, or the entire body of a living creature. This is the reason we can refer to these molecules as the 'book of life.' Technically, this book is called the *genome*; our particular 'book of life' is referred to as the *human genome*.

"The RNA, on the other hand, is an auxiliary molecule, which is actually a family of molecules that help to *translate* the information from the language of the DNA to the language of proteins. Proteins are the final products of the central dogma. These molecules perform so many functions that it would take more than a whole semester to describe all of them. Some of them are the building blocks of many tissues, while others perform a host of different functions, acting like little molecular robots. These robots carry out many processes in our cells such as aiding in food digestion, transporting oxygen from our lungs to the tissues, and transporting waste materials away from the tissues to be expelled from the body. They help to accelerate certain chemical reactions while decelerating or inhibiting others. They are part of the 'motor' that enables us to move our body or parts of our body. Proteins also control and regulate very many processes. These are just a few of the *functions* that proteins perform."

As the professor talked about DNA, RNA, proteins, and the central dogma, Alice wondered why he had not mentioned water — one of the molecules that played such an important role in the previous semester's lessons. But no sooner had she started to mentally construct the question that she wanted to ask then she realized the professor was already giving her an answer.

"You might be wondering why I devoted so much time to teaching you about water during the last semester when all I am talking about now are the three most fundamental molecules at the very core of the central dogma. But don't worry about this, as I will return to the subject of water later on."

"Was he perhaps reading my mind?" a bemused Alice asked herself.

The professor continued, "I do not expect you to fully grasp what I have said today. We shall have plenty of time to discuss each of these molecules, and then we shall combine all of them to construct the so-called central dogma. But you should know that the central dogma is at the heart of molecular biology, and that molecular biology lies at the core of biology — and of course, that biology is at the heart of life, all forms of life, in fact." He glanced at Kofeau as if to look for his little ally's approval and validation, and the monkey seemed to nod softly back, apparently following the whole discussion.

"I deliberately omitted water," he went on, "the most important and most vital molecule. Indeed, if you read about the central dogma in any textbook, you will come across the three main protagonists: the DNA busy with replication and transcription, the RNA taking the role of the translator to produce the proteins, and the proteins doing what we commonly refer to as multitasking. These 'multifunction' proteins assist the DNA in doing its replication and transcription while at the same time helping the RNA to translate the information from the DNA to proteins. They help regulate all these processes, in effect forming a closed circle.

"As in all theater plays, however, the audience only sees the actors. But in reality there are many activities going on behind the scenes — backstage crewmembers that work hand in hand to deliver what we actually experience. Without the behind-the-scenes crew, no play or movie would come to fruition. The 'backstage workers' that make life possible are the water molecules, without which the central dogma would be a 'frozen' dogma, or if you will, a dogma 'on paper,' incapable of translating into life."

The professor paused momentarily, opening a small bottle of water as he was trying to prove his point. He took a sip, put the bottle down on top of the desk and looked over at Kofeau, who was staring at the bottle. The professor knew all too well the "I-am-thirsty-too" look, and he soon had another bottle opened and beckoned the little monkey to come to him. Kofeau quickly got down from the chair and hurriedly grabbed the bottle of water, gulping down the water. After finishing every last drop, he took out a handkerchief from his bag and wiped his mouth, causing the whole room to erupt in laughter again.

"You have just witnessed the importance of water, courtesy of our friend Kofeau," said Professor Holmes, smiling broadly. "In Nature, animals often go to extraordinary lengths in search of water, just as much as they seek out food. Sometimes they will risk their lives to obtain much-needed nourishment, food, and water.

"In a sense, the water molecule is no less important than any one of the players that I mentioned in the central dogma. Water not only provides the background and the medium in which the processes occur, but water molecules actually take an active role in all of the processes that I have mentioned, as well as in many others.

"Thus, my overarching goal for this semester is, firstly, to teach you the central dogma. This is the central enterprise that takes place in our cells. Once you are familiar with all the events that comprise the central dogma, I will show you how water makes its contribution to all of these processes. You will see why water is so essential to the central enterprise, and in fact, to life itself. Today, we shall briefly discuss the main processes that comprise the central dogma."

Professor Holmes pointed at his slide. "In this schematic diagram you see the three main protagonists: DNA, RNA, and proteins. The curved circle that you see around the DNA symbolically signifies the reproduction — that is the

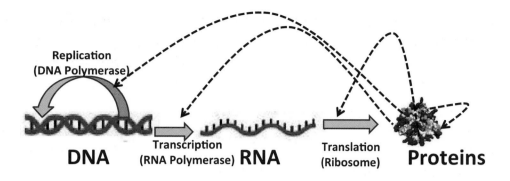

Fig. 5   The central dogma of molecular biology.

repeated *replication* of the same molecule, as long as the cell is alive. The arrow that you see pointing from the DNA to the RNA signifies transmission of information, which is referred to as *transcription*. This means that the same information contained in the DNA is rewritten in a slightly different language on a new molecule, the RNA. The second arrow is the *translation* of the information from the RNA to a new language of proteins, not one but many proteins. Some of the proteins go on to perform various chores that do not belong to the central dogma. This is shown by the multiple arrows on the right-hand side, but you can also see curved arrows that go back to the DNA and the RNA. These signify that some proteins that are *produced* by the schematic central dogma as shown here are actually taking part in the very process that leads to their production.

"This semester we shall discuss the molecules involved in the processes that occur in this schematically described central dogma. In the next class, I will discuss the DNA molecule and its central importance in genetics. You are surely familiar with the fact that offspring 'resemble' their parents, so you are familiar with the phenomenon of *heredity*. The transmission of some traits from ancestors to their descendants is a phenomenon that mankind has recognized from time immemorial. But how it happened was not understood until the structure of DNA was unraveled. After this discovery we now understand many of the details about heredity on a *microscopic* level, whereas for many years these phenomena were only understood on the macroscopic level."

Processing the professor's every word, Alice realized that he was actually echoing what he had discussed last semester. While she had felt completely lost on the first day of last semester, she was now much more assured of her own knowledge, thanks to her frequent visits to the laboratory. In contrast to last semester's unchartered territory, she now felt she was on familiar ground.

The bell rang, signaling that the class was over, and before Alice could even approach Professor Holmes, he made a wild dash towards the door, with Kofeau in hot pursuit. Alice chuckled to herself, wondering what else the professor had up his sleeve.

2

DNA, the Carrier of Genetic Information

Fig. 6 The structure of DNA.

Never to be outdone when it came to punctuality, Alice was the first to come to class the next day. There was a faint musty odor in the room, and she opened the windows to let some fresh air in. The crisp morning carried with it a hint of freshly mowed grass. As the invigorating air started to find its way into her lungs, she started to focus on the day's lesson. The three molecules "at the heart of molecular biology," as Professor Holmes had put it, echoed in her mind.

The assembled class was soon treated to a now familiar sight: a professor clutching his books and a little monkey ambling purposefully behind him. Professor Holmes wasted no time and immediately got down to brass tacks. On the screen was projected an image that resembled a sequence of beads and with what looked like arms emerging from each bead.

"Today we shall focus on one molecule: DNA," announced the professor. "This is an acronym for deoxyribonucleic acid. But don't be intimidated by this complicated sounding term. I will show you the structure of these molecules. You don't need to memorize the full name; just remember that DNA is the molecule that carries genetic information. As I have said, the DNA can be referred to as the 'book of life'. Later, we shall discuss why it makes sense to do so.

"For many years scientists were puzzled about the secrets of heredity, how some features or traits are passed down from parents to their offspring from generation to generation. For example, eye color. In the past, people were mystified by the phenomenon of heredity. What caused the offspring of one animal to be so similar to their parents, yet different in some respects? At the beginning of the 20th century, people speculated about the mechanism of this mysterious phenomenon. They believed that there must be some molecules on which information about the traits of a given animal is 'written,' and transmitted from generation to generation. No one knew, however, what these molecules were, in which 'language' the information was written, and how this information was transmitted. It was a fascinating and intriguing question that preoccupied many scientists.

"Today, we are fortunate to have discovered these molecules. We also have some ideas about the language of the information written on them, and many details of the transmission of this information from generation to generation. Instead of describing the historical development of genetics and the theory of heredity, I will start from what is known today."

Then the professor asked, "How many letters are there in the English alphabet?" Immediately, the class answered in unison "26." Alice was silent, wondering why the professor had asked such a simple and obvious question. "Everyone knows how many letters there are in the English alphabet," she said to herself. She was not thinking about answering the question, but rather why he had asked such a question.

"Correct! There are 26 letters that we can string together to come up with words, sentences, paragraphs, and so on. What emerges is that the sequence of letters might or might not carry some meaningful *information*. I will discuss shortly the kind of information that the DNA carries, but before doing so consider the following 'sentence,'" the professor said, writing on the board:

MODIFIED SENTENCE: **WORROMOT NIAR LLIW TI**

"Do you know what this means?" he asked the puzzled students. "No? You don't see any message in this sentence? What if I rewrite it in the reverse sequence? I get a new sentence." The professor scribbled on the board again:

ORIGINAL SENTENCE: **IT WILL RAIN TOMORROW**

"I will refer to this as the 'original' sentence, and the previous one as a 'modified' sentence. I'm sure you can see that the original sentence conveys some information. Although you did not realize it, it is also true that the modified sentence carried the same information as the original but in a different presentation. All you need to know is the *rule I* used to rewrite the 'information' in the new form. Now look at the following sentences."

CODED SENTENCE: J U X J M M S B J O U P N P S S P X

CODED SENTENCE: K V Y K N N T C K P V Q O Q T T Q Y

"Can you tell what information is carried by these letters? It is easy to see that the first sentence was constructed by replacing each letter in the original sentence — 'It will rain tomorrow' — with the consecutive letters in the alphabet. The second sentence is obtained by replacing each letter by the second consecutive letters in the alphabet. So you see that if you know the rule or the 'code' that I used to convert one sentence to another, you can convert back using the *same code* to retrieve the original sentence.

"Of course, here I used a very simple code, which is easy to work out after making a few guesses, but there are encrypted messages that are extremely difficult to decipher. Indeed, even the original sentence, which all of you could easily read and understand, would be meaningless to someone who does not know any English. However, he would be able to understand it if it were translated into his own language. In this case, it would not be a letter for letter translation but rather word for word, for instance.

"The DNA is written in a *four-letter* language, denoted by the symbols A, C, G, and T. The reason for choosing these particular symbols is not important for now. We'll return to that some other time.

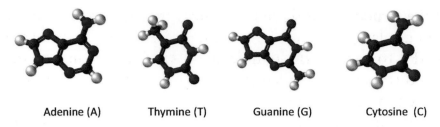

Adenine (A)   Thymine (T)   Guanine (G)   Cytosine (C)

Fig. 7   The four 'Letters' in DNA.

"Consider the chemical formulas of these four symbols. These are not regular 'letters'; each one actually represents quite a complicated molecule. We shall soon simplify the structure of these molecules. We shall eliminate all the details except for one feature that is important to us. With these symbols one can write a phrase like this," said the professor, writing on the board:

TTAGATCACCT

The students turned to one another with questioning looks, and with hushed voices asked "What?" The professor quickly interjected, "Don't try to look for a *meaning* in this sequence. But consider that this sequence does indeed carry some *information*. It is not information in the usual sense that one uses every day, for instance in the English sentence we wrote earlier. In fact, the sequence of letters in the DNA-alphabet carries not one but two kinds of information. Perhaps a better way of saying it is that the information on the DNA is used in two 'directions.' In one 'direction' the information is used for the process of *replication*, whereas in the other it is used for *translation* from DNA language to protein language. Today, we are only discussing the former. We'll discuss the latter in future lessons.

"The *replication* of the DNA is one of the fundamental processes in the central dogma of molecular biology — and is the fundamental process in heredity. Before we discuss the details of replication, take another look at the four molecules represented by the letters A, T, C, and G. I have simplified the structures of these molecules. Two are small molecules (C and T), and two are larger (A and G). Furthermore, I have added some legs (L), and hands (H) to each of these molecules. All that I want you to remember is that a hand (H) can hold a leg (L), and a leg can hold a hand. This means that T and A can hold each other by means of *two* 'bonds,' whereas G and C can form *three* bonds. For us a 'bond' means holding a hand with a leg and vice versa.

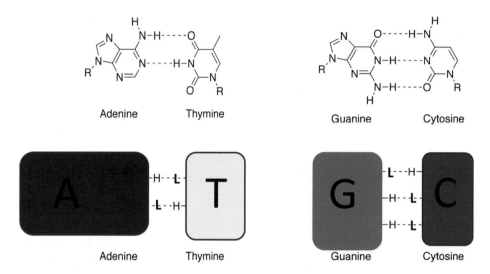

Fig. 8   The pairings of the four bases; A = adenine, T = thymine, G = guanine, C = cytosine.

"To understand the process of *replication*, consider a *code* whereby each letter in the DNA-alphabet is replaced by another letter from the same alphabet according to the following chart, or 'dictionary.'" The professor wrote the following on the board:

$$A \rightarrow T$$
$$C \rightarrow G$$
$$G \rightarrow C$$
$$T \rightarrow A$$

"With this dictionary we can translate, or rather *transcribe*, the sequence we wrote before as follows:

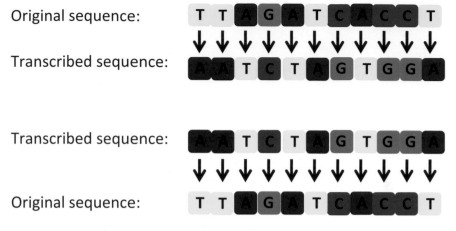

Fig. 9   Transcribing and re-transcribing.

"What we have is a new sequence with the same letters, and of course with the same 'meaning,' whatever the meaning we had in the original sequence. I say *meaning* not in the sense that you understand the word 'meaning,' but in the following sense. Suppose we redo the transcription of the new sequence, using exactly the same 'dictionary' as before, what will we get? What we will get is the exact original sequence. Thus, if the original phrase had a 'meaning,' it must have been hidden in the transcribed phrase, and by transcribing it again we have recovered the original meaning, assuming it had any meaning.

"Are there any questions?" the professor asked rhetorically, knowing very well that there would be none. After all, what he did was translate some seemingly meaningless sequence into a new sequence, and then back to the original sequence.

Suddenly, Kofeau raised his hand, seemingly poised to ask a question. "Well, Kofeau, please allow your classmates to ask the questions," the professor said, giving way to an impish grin. The little monkey shrugged his tiny shoulders and scratched his head, as if resigned to his fate. The students could not help giggling at yet another funny scene. But no one ventured to ask a question, until Alice finally raised her hand.

"Everything that you said makes sense, but what I don't understand is what it has to do with heredity?" said Alice, most of the class nodding their heads in agreement. The professor was ready with an answer, anticipating that they would not fully understand what he had said.

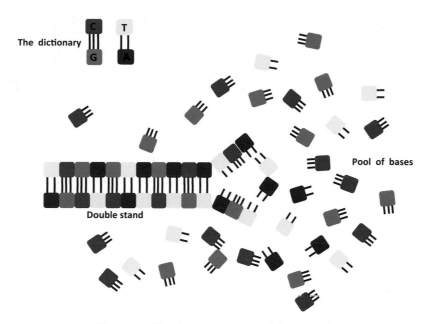

Fig. 10 Unzipping the double strand.

"That is a very good question. I made it clear last semester that if there are things that anyone in the class doesn't understand, please feel free to ask. You are right, Alice! So far I have not explained what actually goes on within the cell. Be patient. Even after the next few lectures, there will be more questions, and hopefully not only from Alice, and I will be glad to clarify and answer all of them.

"Let's go one step further and discuss how this transcription business actually occurs. In the first example of translation — the sentence containing the information 'It will rain tomorrow,' — I gave you the 'code.' The method I used was to replace each letter by another one. In the cell, there is no one actually doing this translation, but instead the process is done by 'holding hands and legs,' in order to form bonds. If you like, you can say that the coding is done 'mechanically': letter A binds to the letter T, and letter C binds to the letter G. Again, I urge you at this stage not to look at the 'meaning' of the message being coded or transcribed.

"Now suppose you have a sequence of letters such as the one we had earlier, and suppose that this sequence is immersed in a liquid in which the single letters A, C, T, and G are floating around. What will happen is that after sometime you will get a double sequence. Remember, A can only be attached to T, and C can only be attached to G. Alternatively, you can remember that blue (A) attaches to yellow (T), and red (C) attaches to green (G). Do you remember the code chart?" said the professor, gesturing at the board. "In the figure you can see the actual chemical formulas of the four bases. The details of these formulas aren't important, so I have replaced the formulas with colored rectangles. What is important is that A binds to T by holding two arms, while 'G' binds to 'C' by holding three arms.

"For now let's not concern ourselves with whatever the information is in the original sequence. It is now in the form of an original and its encoded copy, like an image and its negative, but now the 'original' and the 'negative' are bound to each other. This is referred to as the double strand. Don't worry, for the moment, about the 'meaning' of the information. Simply view the 'original' sequence we started with as a message and the 'negative' as the same message but encoded by the dictionary that we had earlier.

"Now I shall describe the extraordinary process of *replication*. Suppose you start with the 'original' and 'negative' pair. This is the double strand immersed in a solution of randomly floating letters A, C, T, and G. Next, suppose that at some signal, I start to 'unzip' this pair sequence — something like this," said the professor, pointing to his next slide. "What do you think will happen?"

Alice could see the answer but was not sure, so she decided not to raise her hand. Some students did, however, and one said, "Well, I guess that the two open parts of the zipper, like you see in this figure, will have free arms exposed to the liquid. Since there are many letters in the liquid searching for free arms, any free arm in the open part of the zipper will bind to the corresponding arm of one of the letters in the liquid."

"That is basically correct," the professor acknowledged with a smile. "Of course, in reality the process is more complicated, but for our purposes this is indeed what happens. You can imagine this as the double sequence that starts to open. For the first pair that is unzipped, in our case A and T will be wide open, with free A-arms and free T-arms. Since there are plenty of As and Ts in the solution, the A-arms of the unzipped end of the sequence will bind to the T from the solution, and the T-arms of the unzipped end will bind to an A from the solution.

"You can easily visualize this if you continue this process. As the original 'double sequence' is unzipped, two new zippers are formed. What is more interesting is that the two new zippers that are formed are exact replicas of the original zipper, as shown in this figure. I said *exact* replicas. Indeed, the process I have described produces an exact replica of the original double strand. In reality, some errors might occur in this replication. However, these errors are rare — quite rare, in fact — but important in the theory of evolution."

While the professor was talking, an animated clip flashed on the screen. A zipper opened up into two separate strands, and at the same time each of the strands became a double strand by accumulating letters, or 'teeth' from the environment.

"The double strand we talked about is only a small part of the DNA," the professor continued. "It is referred to as the double helix because of its special structure. The entire process of unzipping and then forming two 'offspring'

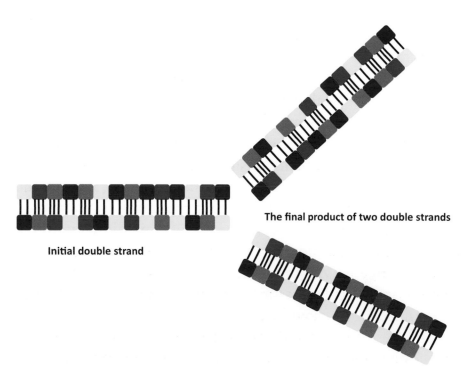

The final product of two double strands

Initial double strand

Fig. 11   The final product of two double strands.

or new zippers is called *replication*. I hope you now see how the 'message' in the DNA replicates itself. This is only one process in the central dogma.

"The structure of the double-helix model of DNA was discovered by James Watson and Francis Crick in 1953. This discovery immediately suggested a mechanism for storing genetic information, copying this information and producing multiple copies. This is considered one of the most important discoveries in biology, and it revolutionized the whole field of biology and genetics.

"As I said earlier, the process I have described to you is the *bare* process of replication. In the actual cell there are many 'agents' that are involved at each step of this remarkable process. For today we can summarize that we have learned one element of the central dogma. We started with a sequence of letters, which we will refer to as a *single strand*. A single strand together with its 'negative' strand forms the *double strand*. The double strand can replicate itself to form two identical *offspring*.

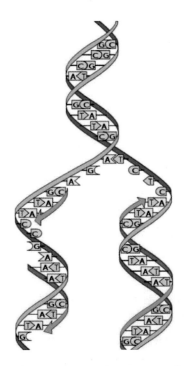

"This process can go on and on provided there is an unlimited supply of A, C, T, and G, and of course the machinery that orchestrates all these processes. The main agents that execute all these processes are the proteins. In the next class we'll talk about proteins, another element in the central dogma. Then we'll discuss the relationship between these two elements. Let's call it a day," the professor added, suddenly looking tired.

On her way to the next class, Alice wondered how this replication of the DNA is 'translated' into the replication process of a whole animal. After all, an animal is not a single strand of letters multiplying by zipping and unzipping. But she clearly remembered Professor Holmes' last words: that this was only one *element* of a bigger picture, like a single piece in a jigsaw puzzle. She knew that after a few more classes everything would be clearer, but there was always the option of paying a visit to the professor's lab...

Fig. 12   Replication of DNA.

3

Proteins: Multipurpose, Multifunctional Molecules

The dimming of the lights not only signaled the beginning of the class, but it also created a conducive atmosphere for Kofeau to doze off. Unlike the previous days when he came to class oozing with energy, today he was lethargic and had napped all the way through the first half of the lecture. Alice, on the other hand, was wide awake, and was waiting for an opportune moment to ask Professor Holmes some questions she had carefully formulated in her mind.

Although she had understood everything in the previous lecture, she could hardly wait for the next part of the story: how the replication of a single strand of DNA explained the hereditary traits of specific animals, what kind of information was carried by the DNA molecule, and if there were any agents that could read this information and execute the instructions that were 'written' on the DNA. She had so many questions and was bursting to ask them.

Once again, the central dogma slide appeared on the screen. This time, though, the professor's laser pointer was no longer indicating the DNA; it had moved over to the protein side. He cleared his throat and began.

"In our last class, we learned about one aspect of the central dogma: the DNA molecule and the mechanics of its replication. Today our discussion will focus on another important player in the central dogma, the proteins. Once we familiarize ourselves with these versatile, multipurpose molecules, we shall combine the two parts of the dogma — in other words, how the 'information' embedded in the DNA is translated into the language of proteins.

"In the DNA, proteins may be viewed as a sequence of letters. Just like the DNA, this sequence carries some 'information,' but now the 'information' is written in a new language with a different alphabet, more specifically an alphabet consisting of 20 letters. For simplicity let us just take the first 20 letters of the English alphabet:

a b c d e f g h i j k l m n o p q r s t

"I should mention that in biochemistry the symbols or letters used for the different amino acids are different from the first 20 letters of the English alphabet. In the next slide we shall see the actual structures of the amino acids.

"For our purposes we do not need to memorize any of this. What matters at this time is that a protein may be viewed as a sequence of beads, but instead of four kinds of beads represented by the capital letters A, C, T, and G in the case of the DNA, we shall label the various beads of the protein with the different letters of the English alphabet."

Professor Holmes turned to the polystyrene container he had brought to class, flipped open the lid and took out a still-frozen chunk of beef.

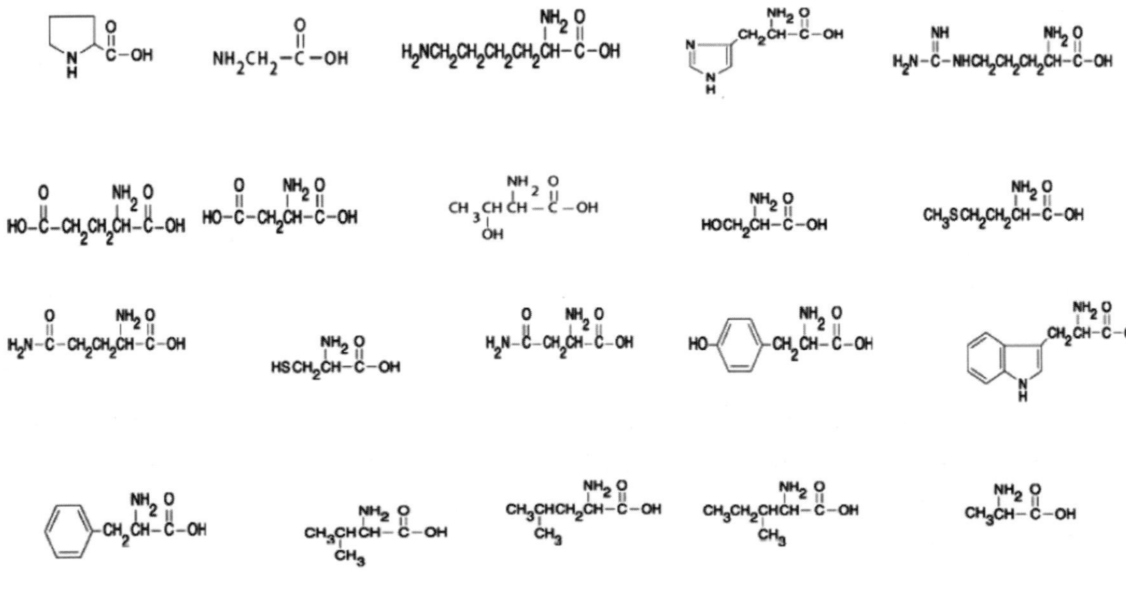

Fig. 13 The 20 amino acids.

"This piece of meat is mostly made up of protein," he declared, holding it up for the class to see. "Imagine sinking your teeth into a tender and succulent slice of tenderloin steak."

The half-awake Kofeau suddenly opened his eyes, stood to attention and started sniffing the air. Minutes earlier in a sleepy daze, he now seemed to be wearing a broad smile on his little face.

"Hmmm…,"said the professor as he glanced at Kofeau. "Someone seems to have woken from a dream! Well, aside from meat, we also love pasta, pastries and candy, all of which are mainly carbohydrates. And what about you, Kofeau? What do you like best to eat?" Without hesitation, Kofeau pointed to the slab of beef lying on the tabletop, and the delighted students cheered and clapped enthusiastically.

The professor waited for the applause to die down. "But seriously, what are proteins? Basically, each protein is a string of amino acids. There are 20 different amino acids represented by 20 letters of the English alphabet. A typical protein could have 100 to 150 amino acids. Can you imagine how many different proteins, say of 100 amino acids, are possible? The answer is simple, but the number is mind-boggling: at the first position we can put one of the 20 amino acids, and again at the second position we can put one of the 20 amino acids, and so on until we reach position 100. So altogether we have:

$$20 \times 20 \times 20 \times \cdots \times 20 = 20^{100} \approx 10^{300}.$$

"This is an unimaginably large number — one followed by 130 zeros! It's the number of possible proteins consisting of 100 amino acids; in principle, we can imagine many more proteins of different sizes. The number of proteins that we actually have in our bodies is only a tiny fraction of all possible proteins and yet it is still quite a large number. At this point I would suggest that you do a small exercise. Suppose for simplicity that we write at random a three-letter word in English. How many 'words' are possible? There are:

$$26 \times 26 \times 26 = 17,576.$$

"This is quite a large number. Now write some of these 'words' explicitly: *aaa*, *aab*, *aac*, *aad*... Can you imagine how many of these 17,576 'words' have *meanings* in English? For instance: *eat*, *bit*, *kit*, *bat*... Clearly, there are fewer words with meanings.

"Now let us go back to proteins. In the world of proteins we have really huge numbers of sequences or 'words' of size 100. Actually, we have 20^{100} of them. But only very few have 'meanings.' Here, *meaning* means that a particular protein had some advantage, and therefore survived the very long process of evolution. Thus, from an incredibly large number of possible sequences only a tiny fraction of sequences are now in use in our cells. Note that a tiny fraction of 20^{100} could still be a large number, say ten thousand, a hundred thousand or even millions — which is still a small fraction of 20^{100}.

"What are the *advantages* — or the 'meaning' of the existing proteins?" Professor Holmes paused momentarily to determine whether his class was following. A student who was seated in the back row raised his hand and said, "Proteins are one of the basic foods that form part of our diet..."

The professor took out a throat lozenge and removed its wrapper, and the crackling sound made Kofeau look up. Then he produced a small pack of multicolored candies from his pocket, and waved them at Kofeau.

"So what do you prefer, Kofeau, proteins or candies?" The little monkey pointed to the candies, and in no time at all, he was standing beside the desk and picking them gently from the professor's outstretched hand. Returning to his seat, he started to devour the bounty, and as soon as he had finished eating, he collected the empty wrappers together, tucked them inside his bag and carefully wiped his hands.

"You are correct," said the professor, addressing the student in the back row. "Proteins are indeed an important component in our diet. But what I meant was: What do they do in our bodies? Perhaps I should explain in response to the answer I just heard. The proteins we take as part of our food intake is digested in our bodies, while the proteins *needed* by the body are synthesized according to the 'instructions' written on the DNA.

I shall discuss the meaning of these 'instructions' in a later class. At this stage I want to discuss proteins that exist in our bodies, and not how they are created or produced by the body, nor what they actually do in our cells.

"Remember that DNA is a molecule which carries 'information.' This information is used either for the purpose of *replication*, or for *translation* into proteins. We shall learn more about this in the next class as well. These are basically the two functions of the DNA.

"The proteins, on the other hand, are far more versatile and have a multitude of functions. Proteins are the building blocks of many tissues. They are also the builders of larger structures from the smaller building blocks. For instance, proteins are assembled into larger units that are the parts of the muscles. Proteins are important carriers of smaller molecules. The most important case is the hemoglobin, which transports oxygen from the lungs to the cells and transports back the carbon dioxide from the cells to the lungs. Proteins also function as enzymes. They catalyze or accelerate numerous chemical processes in the cells. It is no accident that proteins are referred to as 'molecular robots.' We have mentioned only a few of their functions. There are many more that we know about — and perhaps a lot more that we are yet to understand.

"I should mention that proteins also play a crucial role in the process of DNA replication, as well as in the process of translating the information 'written' on the DNA into the language of proteins. These functions are part of the central dogma that we have been studying in these classes. Many of the functions of proteins are 'outside' or 'do not belong to' the central dogma. We shall return to some of these functions after we have grasped the essence of the central dogma."

For most of the class, Alice had been quiet. She had no difficulty following what the professor had said. No difficulties and no questions. But she was eager to visit the laboratory again and was too shy to ask the professor. Clarifying certain points in the class would be the perfect excuse for her to visit the professor in his lab, and the thought made her smile.

On her way home, Alice went over the new things she had learned. She felt that she had understood quite well the role of the DNA as a replicator. She had some idea of what proteins were, though only a faint idea of what proteins did or how they *functioned*, as the professor had put it, and she had absolutely no idea how the 'information' contained in the DNA was translated into proteins. She was looking forward to learning about the third element in the central dogma, and how the three elements were linked together.

4

Translation of Information from DNA Language to Protein Language

"Good morning, my young friends!" Professor Holmes greeted his class warmly. The professor's unusually informal and friendly demeanor pleasantly surprised the students. The change in the professor's mood had just started to sink in when he started talking about the puzzle of the central dogma. "Back to normal," thought Alice.

"We know what DNA is, yes? We also know what proteins are. Now our task is to connect the two, very much like connecting the dots, so that in the end we will be able to see the big picture," he said cheerfully.

"Before undertaking the task of linking the DNA to proteins, let me describe the link metaphorically. As an example let's consider a book, which we all know has different chapters. Let's assume that the book deals with the construction of a building. One chapter contains information on how to make bricks, say, while another one deals with tile making, and a third focuses on making windows and doors, and so on. There is also a chapter dealing with how to make the floors, the walls and the roofs, and so forth.

"In addition, and take note — this is an important addition you will not find in a regular book — the book contains chapters describing the various workers and their responsibilities. There are also chapters dealing with robots specializing in bricklaying, tile setting, and carpentry; with robots moving construction materials from one place to another; with robots acting like foremen and giving tasks to the workers. Some robots regulate and control the work, commanding others when to do certain things, and when to stop. Some accelerate various processes, and some decelerate them, and the processes go on and on.

"You can imagine that there's a huge volume of information in this book. Listing all that information would be a tedious process. If there were an agent or a translator who performed the task of reading, sifting, processing and translating all this information, executing the information would be considerably simpler. The most important thing that you should realize is that the book also contains information on how to *read* the information in the book — and how to *execute* the instructions.

"With a regular instruction manual, one implicitly assumes that a person or an agent will read and execute the instructions, right? What makes the DNA instruction manual so extraordinary is that it contains not only instructions on how to construct certain things, but also on how to construct the *agents* that will read and execute all the instructions. This is what differentiates a regular book from this very special 'book of life.'" The professor paused to allow the students to absorb what he had just said.

"Before we continue, let's take a ten-minute break. Some of you seem to have wandered off to Never Never Land," the professor added with a big grin. With what seemed like a collective sigh of relief, the students got up one by one to stretch their legs, some chatting quietly and others checking their mobile phones.

"I see that you are all recharged," said the professor hopefully, as the short break came to an end. "Let's get back to business — back to the DNA and its connection with the proteins. As you can all see in the diagram that we have here, there are three main processes involved in the central dogma. It is often said that the 'information' contained in the DNA *flows* from the DNA to the RNA, and from the RNA to the proteins. We'll discuss the meaning of this 'information flow' after we describe the processes themselves. Then we'll be able to understand in what sense the concept of 'information' is used in the central dogma, as well as in the statement that DNA contains the *genetic* information that is passed from generation to generation.

"We shall start with a sequence of beads representing the letters A, C, T, and G. I presume you still recall how this sequence can copy itself. So we can say that whatever information this sequence contains, it is reproduced in the process of replication. If there are no errors in this copying process then the information is exactly replicated. This is symbolized by a curved arrow in the figure.

"Next, we move on to the more complicated process of *translation*. The first step in the process is called *transcription*. Imagine that the sequence of letters A, C, T, and G may be transcribed into another sequence of four letters, but this time of different beads which we shall denote A′, C′, T′, and G′. The coding rules are almost the same as in the replication of DNA, but now we have a new 'dictionary':

$$A \Leftrightarrow T'$$
$$C \Leftrightarrow G'$$
$$T \Leftrightarrow A'$$
$$G \Leftrightarrow C'$$

"For instance, the sequence we had earlier will be transcribed into a new sequence as follows:

Original sequence: T   T   A   G   A   T   C   A   C   C

                            ↓   ↓   ↓   ↓   ↓   ↓   ↓   ↓   ↓   ↓

Transcribed sequence: A′ A′ T′ C′ T′ A′ G′ T′ G′ G′

"I should note, however, that the letter T′ stands for a new base called uracil (U), but for simplicity we have used the letters A′, C′, T′, and G′ for the transcription to RNA. You can see that uracil is very similar in structure to thymine.

"Again, without specifying the kind of information contained in the original sequence, the transcription process produces a new sequence that contains the same information but in an encoded form. This new sequence is called

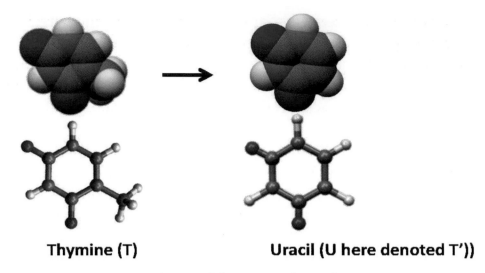

Thymine (T)   **Uracil (U here denoted T'))**

Fig. 14 Thymine and uracil.

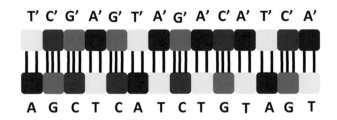

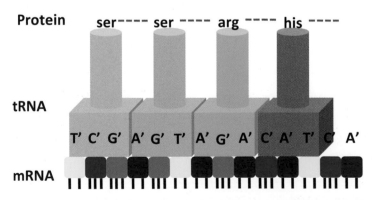

Fig. 15 Transcription from DNA to mRNA, then translation to proteins.

messenger RNA, or *mRNA*. I should mention that in some viruses, there is a flow of information from the mRNA to DNA using the same dictionary, but we shall not discuss this here. The process of transcription is different from the process of replication, just as taking a photograph of a page gives us a *negative* of the page, whereas photocopying gives us an exact replica of the original page. Similarly, if we transcribe the page using a code, we will have the same information, whatever the information is, but in an encoded form.

"We now move on to the next step whereby the information contained in the RNA — the information that originated in the DNA — is translated into a new language with a new alphabet. You will remember that a protein is a sequence of amino acids. Clearly, we cannot encode an alphabet of four letters into one of 20 letters using a one-to-one dictionary. A sequence of amino acids like

$$h - a - e - b - n - q - t$$

cannot be produced by a four-letter alphabet with a one-to-one code.

"Let us try pairs of letters in the DNA or the RNA to encode 20 letters in the proteins. We have $4 \times 4 = 16$ different pairs of letters in this case, and clearly this is not enough to code the 20 letters of the protein. With triplets of letters we have $4 \times 4 \times 4 = 64$ different triplets, and this is clearly more than is needed. It turns out that Nature uses the triplet of letters in the RNA alphabet to code for all the 20 letters of the protein. Look at the 'dictionary' for the translation from a three-letter codon to an amino acid. Such a code is said to be degenerate. There are several triplets referred to as a 'codon' which translate to one amino acid, for instance

codon = C' G' $C' \to$ (*arginine*), represented by the letter r, or R in the diagram.

"Of course, you do not need to memorize this 'dictionary.' Just have a look at it and be impressesed by what Nature has achieved — and how much scientific effort was invested in deciphering this code. There are several codons that encode to a 'stop sign' rather than to an amino acid. This is equivalent to a period in a sentence. This 'dictionary' is called the *genetic code*."

Seeing the students scrambling to copy everything, the professor repeated, "You don't need to copy or memorize this table. It is enough for now to remember that a sequence of letters in the DNA is transcribed to a new sequence of letters in the mRNA, and this sequence is translated into a sequence of amino acids. Of course, this is a very schematic description of translation. In fact, the actual processes involved in the translation are very complicated. Many scientists have made an effort to figure out the entire puzzle. I hope you can imagine how much work and effort have been invested in this endeavor, and that a great deal of ingenuity was required to construct

The Genetic code
RNA codon table

nonpolar	polar	basic	acidic	(stop codon)

Standard genetic code

1st base	2nd base U		C		A		G		3rd base
	UUU	(Phe/F) Phenylalanine	UCU		UAU	(Tyr/Y) Tyrosine	UGU	(Cys/C) Cysteine	U
	UUC		UCC		UAC		UGC		C
U	UUA		UCA	(Ser/S) Serine	UAA	Stop (*Ochre*)	UGA	Stop (*Opal*)	A
	UUG	(Leu/L) Leucine	UCG		UAG	Stop (*Amber*)	UGG	(Trp/W) Tryptophan	G
	CUU		CCU		CAU	(His/H) Histidine	CGU		U
	CUC		CCC		CAC		CGC		C
C	CUA		CCA	(Pro/P) Proline	CAA	(Gln/Q) Glutamine	CGA	(Arg/R) Arginine	A
	CUG		CCG		CAG		CGG		G
	AUU	(Ile/I) Isoleucine	ACU		AAU	(Asn/N) Asparagine	AGU	(Ser/S) Serine	U
	AUC		ACC		AAC		AGC		C
A	AUA		ACA	(Thr/T) Threonine	AAA		AGA		A
	AUG*	(Met/M) Methionine	ACG		AAG	(Lys/K) Lysine	AGG	(Arg/R) Arginine	G
	GUU		GCU		GAU	(Asp/D) Aspartic acid	GGU		U
	GUC		GCC		GAC		GGC		C
G	GUA	(Val/V) Valine	GCA	(Ala/A) Alanine	GAA	(Glu/E) Glutamic acid	GGA	(Gly/G) Glycine	A
	GUG		GCG		GAG		GGG		G

Fig. 16   The genetic code.

this theory — a beautiful picture made up of hundreds or thousands of small pieces as if it were a mosaic."

The professor paused and cleared his throat, gauging whether he had successfully conveyed to the class his own excitement and admiration of this extraordinary scientific achievement. "So," he continued, "at this stage we have a full description of the central dogma. What remains to be clarified is the meaning of the term 'information' flowing from RNA to proteins."

Sensing that he had bombarded the class with an awful lot of information, the professor paused for a moment and asked, "Before I continue with the issue of information, I'm just wondering if there might be any questions so far?"

It was so quiet you could have heard a pin drop. Kofeau, who appeared to be afraid of being asked, avoided the probing eyes of the professor by looking the other way.

"So far so good then?" Again, no one said a word.

Professor Holmes knew that no one could have understood the meaning of the information that was contained in the DNA and then translated into proteins, but that was exactly what he intended to tackle next. For her part, Alice was quiet because she felt that she hadn't missed any of the points that the professor had made. She just wanted to understand what precise form the information took, and she had hoped that day's class would shed light on the matter.

"Let me assume then that everything is clear so far," continued the professor with a smile. "Therefore, I will move on to the more difficult issue of the meaning of information contained in the DNA.

"You will recall that I used a book as a metaphor. I presume that you all understood what kind of information that book contained. It had very detailed information regarding the production of various construction materials as well as a section on robotics — those little robots that combine, assemble, control and regulate all the processes.

"By comparison, the information contained in the DNA is more abstract. It is not written in any language you have ever heard of, yet people say that the DNA contains all the information needed to make a specific animal. A dog will have one book, a cat will have another book, and we human beings will also have a different book. All of these books are similar in some sense, but they are different. For instance, many of the proteins that we have in our cells are also found in bacteria. Also, the processes of replication, transcription and translation are similar in all living creatures. Yet differences exist. We differ from bacteria, but we also differ from dogs and cats — and as you very well know, there is not much difference between monkeys and us.

"You can think of the difference as the difference between a book with instructions on how to make a small hut and another book for a high rise building. Both must have instructions on how to make the cement, the bricks, the tiles, and so on. There are also instructions on how to arrange the bricks and cement them together, how to assemble the roof, and so on and so forth. However, although there may be common information in these books, they are still different in other ways. By the same token, creating an amoeba is different from creating a dog or a cat, and they are only remotely similar to creating a man or a monkey."

Suddenly, as if to illustrate the professor's point, Kofeau took out some measuring spoons and cups from his bag and started measuring out imaginary ingredients. The class was immediately in stitches again. Alice, who also found Kofeau's performance hilarious, wondered how the little monkey's book differed from her own — or perhaps the professor's.

"Well, I can tell from your reaction to our friend Kofeau's demonstration that books about making cats and dogs appeal to you," said the professor, grinning. "I meant that metaphorically, of course, as they are rather different from recipe books. Neither dogs nor cats, nor any of us here have such 'do-it-yourself' instruction manuals!" Once again, the class roared with laughter.

The professor tapped his pen on the desk and the noise quickly died down. "The DNA is a *meaningless* sequence of letters. This sequence is transcribed to the RNA, which serves as an intermediate agent carrying whatever information is in the sequence from the DNA into protein. The RNA is threaded into a protein-production machine called the *ribosome*, which synthesizes proteins. Thus, the sequence of RNA enters into the ribosome on one side, and proteins are produced on the other side. It's a lot like when a magnetic tape, which has a sequence of tiny magnets, is inserted into a cassette player, where the sequence of 'magnetic letters' is translated into musical tones. Similarly, the sequence of 'magnetic letters' on a video cassette is translated by a VCR into a series of pictures, which we call a film.

"In the same vein, an RNA entering the ribosome is translated into a specific protein. The process is very complicated, and we will not discuss it here. Once a protein is released, it either acts alone or with a combination of other chemicals available around it to do certain things. On the other hand, some proteins simply aggregate to form larger multi-protein units, while some help

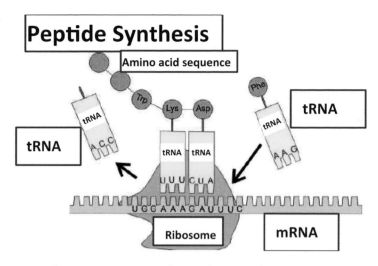

Fig. 17 Schematic process of translation of mRNA into proteins.

other chemical reactions in forming new compounds. These are the little robots that execute a variety of functions. Eventually, a whole cell is created through the concerted efforts of all the robots using all the raw materials available. The process of creating a cell is quite complicated and the process does not end there. Cells of different forms are created by different proteins, which in turn originate from different segments on the DNA. We shall return to that in a moment. Once all the different cells are created, they combine to produce a multicellular structure and eventually the whole animal is formed.

"As in a book on constructing a house, which has different parts, the DNA book also contains different chapters pertaining to the different parts of the body, and also different information which is specific to different cells. It is interesting to note that each cell contains the entire DNA book. However, each cell 'reads' and translates only the chapter that is relevant to that specific cell, for instance, a nerve cell, a red blood cell, a liver cell or any other cell.

"Thus, we can say that the information contained in the DNA is translated into proteins, the proteins in turn assemble many structures within the cells, and the cells combine to form organs and eventually the whole animal. In this sense, we can say that the 'information' contained in the DNA is the information about the specific animal. It goes without saying that the whole process needs both raw materials such as oxygen, water, minerals, and so on, as well as energy. But these indispensable components do not contain the information about the animal.

"We shall learn in the course of time the importance of water in the processes that involve proteins, and that the water molecule acts as if it has four arms — pretty much like Kofeau here who has four arms." Somewhat contradicting what he had just said, he added, "Well not exactly four arms, but two arms and two legs. However, we can ignore the difference between the functions of an arm and a leg…"

Alice was perplexed. She had been following everything that had been discussed in the class. She understood what it meant to have information flowing from one place to another. She could visualize a book containing information: she imagined taking a book from a store, taking it home and reading it, and picturing how the information flowed from the book to her brain. But there was a missing link. What was the "agent" that "read" the information? If the information was a series of instructions like a recipe, where was the agent that executed this information? Alice knew it was time to revisit the laboratory and consult the professor. She grew excited at the thought of actually exploring the world of the central dogma from "inside." The chance of "seeing" who reads what, and who "executes" what, would surely make things clearer.

While Alice had sat lost in thought, the professor had already dismissed the class. She glanced at her watch and realized she only had a few minutes left to get to her next class. She quickly collected her things together and dashed out of the room.

5

Alice Revisits Professor Holmes' Laboratory

Alice was determined to go see Professor Holmes in his laboratory as she missed their one-on-one tutorials. Although she had not really figured out the exact questions to ask, she somehow knew that it had something to do with the 'information' that the DNA carried, what the nature of the information was, in what language it was written on the DNA, what was the agent that read and understood this information and then executed the instructions in order to build proteins. She suddenly realized she had a lot of questions that needed answers.

As she tried to formulate her questions, it dawned on her that the most mysterious or perhaps the most puzzling thing that she had heard in the last class was that information flowed from the DNA to the RNA. While she had followed the concept of information flowing from the RNA to the proteins, the professor had also said that the proteins actually participated in the flow of information from the DNA to the RNA, and to the proteins. To her that sounded strange, as she could not reconcile the fact that information 'flowed' with the help of proteins before proteins were even produced, and the fact that proteins were produced by a production line that was itself made of proteins. It seemed to be a never-ending loop: proteins translated the information in the DNA to produce proteins, which in turn translated information to produce proteins, which in turn translated… Where did the cycle start and where did it end?

One thing was clear to Alice: She needed to get to the laboratory as fast as she could — before she lost her train of thought and forgot all her questions!

The portrait of the professor's daughter greeted Alice as she entered the laboratory, hanging in the exact same spot as when she first laid eyes on it. But as she passed the foyer, she saw something hanging prominently on the divider leading up to the professor's office, something she had never seen before. As she got closer, she saw it was a framed picture of a strange-looking creature with writings on its forehead. With knitted brows she tried to figure out who or what it was, for although it resembled a human being, it was clearly something else. She didn't notice that the professor had approached her softly, and was observing her reaction to the curious figure.

Fig. 18 Alice with professor Holmes.

The professor greeted her cheerfully and introduced the creature. "This is Golem. I bought this picture during my last visit to Prague. I had often wondered if one could ever 'synthesize' or create a man, which is similar to Golem's story. Would that man acquire his own will, personality or ambition, and could he use these to conspire against men? In case you are not familiar with the story, in Jewish folklore, a golem is something created entirely from inanimate matter but some human characteristics can be attributed to it. You may want to read about it if you wish..." The professor trailed off, but then continued before Alice could utter a word.

"Now let me guess... I'm thinking that what we discussed in the last class wasn't completely clear to you, or anyone else for that matter... Sometimes students are afraid or ashamed to ask the professor because they don't want to look foolish... At some point I sensed that your mind had wandered off. Information overload, eh? But like I said, the door to my laboratory is always open to all of my students who may have additional questions."

Alice, who had been feeling quite apprehensive, began to relax when she heard the professor's reassurance that he would always be willing to help.

"I had a feeling that the matter of the nature of the information flow was not really clear to you." Alice nodded shyly, relieved that the professor had taken the words right out of her mouth.

"Well," the professor continued, "the nature of the 'information' involved in the central dogma isn't easy to explain. 'Information' is a very general concept. It isn't easy to define it, even though we feel that we know what it means. Usually, we read something in a book or in a newspaper, or watch something on TV, and we know that we have received some information. It could be information on the weather, information on some crime committed, or even information about specific topics like medicine, music, and arts. The information that we obtain from different forms of media is practically limitless. However, the kind of information that flows from the DNA to the RNA to the protein is not *explicit* information that one hears or sees. It is a more abstract kind of information, or perhaps only 'information' in a metaphoric sense."

The professor paused to let his words take effect. "Let me explain with a very simple example," he said. "What is your favorite cookie recipe?"

Alice's mind was filled with all her favorite recipes for cookies, but before she got the chance to think of the English translation of *lenguas de gato*, the professor started to talk.

"A typical recipe is a list of ingredients and instructions; this is self-explanatory. If you follow the instructions carefully, you will most likely enjoy your freshly baked cookies in no time at all."

Freshly baked cookies. Alice's mouth had begun to water just hearing the words. She could almost smell and taste them.

"You might have not just one but several favorite cookie recipes, as well as food recipes, am I right?" asked Professor Holmes, and at that point Alice acknowledged his question by nodding her head.

"Well," he continued. "I'm sure you have no difficulty in saying that each of the recipes contains *information* very specific to a certain kind of cookie or dish. However, the 'recipe' that is written on the DNA is a little more complicated. Imagine that instead of a recipe that has instructions on how to bake cookies, you also have instructions on how to make a little robot that will read the recipe and will do the entire baking process for you from start to finish, and all you have to do is wait to be served.

"Let us allow more abstractions. You have a book, a tape and a string of letters that you cannot read. Whenever the robot receives a signal — and provided that there are plenty of raw materials around — it starts to work. The robot takes some of the raw materials and mixes them, transforms them to other compounds, and also produces other little robots that help to do all these chores including the reading and executing of the instructions in the book. Whether the task is simple or complicated, you can easily say that all the 'information' required is in the book, whether or not you can read and understand it, because you are confident that the information is contained in the book.

"Likewise, the 'information' contained in DNA is not written in a language we can read. However, we know that when it is in the right environment, and there are some raw materials like water, minerals, and so on, upon giving some initial signal, the DNA starts working, with the help of many molecules in its surroundings including proteins, minerals, water and other 'ingredients.' Besides replication, which is intended to preserve an exact copy of itself, the DNA is transcribed into a new copy we call the RNA. This RNA is then threaded through production machines, which we call ribosomes, and on the other side proteins are produced. Note that rather than a single protein, many kinds of proteins are produced — some of which are like the dough you have for your cookies. The process also produces robot-like proteins that help mix the ingredients and transport some of them from place to place. Some processes are accelerated or decelerated until a whole cell is built, surrounded by protecting walls that we call *cell membranes*. These cells combine to form higher organs and eventually a specific 'cookie' is formed. In this case, the 'cookie' could be an insect, a bird or even a human. Without doubt, it's a magnificent achievement.

"We know many but not all the details of this process. Nevertheless, we should have no issue with saying that all the information required to make a specific animal is there in the DNA. We cannot read it, we cannot hear it, but we observe how it works. Therefore, we deduce there must be 'some agent' that can read and execute all the instructions that are 'written' in this magnificent recipe. So, the DNA can be more appropriately described as a 'recipe for creating an animal,' rather than the more common depiction as a 'book of life.'"

Professor Holmes beamed with satisfaction as he concluded his exposition, and Alice was equally happy that she was beginning to find answers to her questions. But the professor knew he had to expand his explanation a little more so that the concept was crystal clear to his student.

"I hope you have a clearer idea of the vaguely defined concept of information as used in connection with the central dogma," he went on. "Indeed, the term 'information' in this context is not well defined. It is only a way of speaking, a metaphor. You can understand the workings of the central dogma without even using the term 'information.' Think about those long paper rolls you find in a player piano, a self-playing piano. You could say that the paper roll contains some 'information' that is translated into musical notes. However, you could also say that the roll is threaded through a slot in the piano, and each perforation activates one of the hammers, which then strikes a piano string, thereby producing musical notes. So, I don't *have* to use the term 'information' when I describe how a player piano works. In the same manner, I could describe what happens in the central dogma without mentioning the word 'information.' I can simply describe how the DNA replicates itself, how it is transcribed into RNA, how the RNA enters into a machine that produces proteins, and how these proteins do many tasks in the cell, including helping the process of replication, transcription, and translation; you can see that the proteins help the processes that lead to their own production.

"Now that you understand the sense in which the term 'information' is used in connection with the central dogma, I want to draw your attention to a more profound riddle associated with it. Did it ever cross your mind that the central dogma is a kind of 'catch-22' situation?"

Alice smiled because that was exactly what she was thinking about on the way to the laboratory.

"The problem is not really the meaning of the information but how the whole machine works. We know that replication, transcription, and translation are processes that are aided by proteins, but proteins are produced at the end of the production line. So how does this whole process start in the first place? This is the real mystery that is not fully understood. It is similar to the question of what came first, the chicken or the egg. In the central dogma, we can say that the DNA produces proteins, and proteins produce the DNA. But which one came first? This question is one of the mysteries of the origin of life. The DNA is very good at replicating itself, but it cannot serve as a robot. On the other hand, proteins are very good at being robots, performing many functions, but they do not 'know' how to replicate themselves.

"One conjecture that has supportive evidence is the following. In most animals, information flows from the DNA to the RNA and then to the proteins. However, in some viruses, known as retroviruses, there is a flow of information from the RNA to the DNA. This finding suggests a possible solution to the riddle of who came first.

"Perhaps, in the early stages of evolution there was a primitive RNA molecule that served both as self-replicating information carrier, on one hand, and robot, on the other, i.e., it also had enzymatic activity like the proteins.

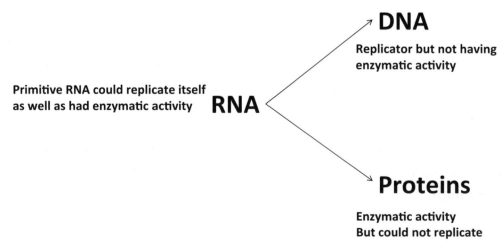

Fig. 19 Possible origin of both DNA and proteins.

The 'personality' of this primitive RNA, in the course of evolution, 'split' in two directions: one evolved into DNA, and the other evolved into proteins. Later, the RNA evolved to perform the translation function while the DNA and proteins took over their specialized functions. This sounds like a logical possibility. However, we might never know the real answer since we were not present when it occurred and we cannot perform experiments to prove that conjecture. Such processes take billions of years to evolve. Still, it is an interesting challenge to conjecture about the origin of the central dogma."

Alice suddenly caught the sadness in the professor's eyes as he looked at the picture of his daughter hanging on the wall. For a moment he seemed to be transported somewhere else.

"Since you took time out to clarify certain issues," he went on, quickly regaining his lively demeanor, "let me assure you that things will become clearer after our next class. I will be talking about proteins, how they are folded into a precise structure and how they are associated to form large structures. In all these processes, there is another important protagonist that plays a crucial role. You know what that protagonist is called, don't you?"

The professor could tell from Alice's smile that she knew the answer, and she beamed even more seeing that the professor was aware of it. Neither spoke. They let the question hover between them, savoring the precious moment.

On the way home, Alice recounted all the things the professor had discussed, trying hard not to miss out a single detail. She knew what the protagonist was in all the processes taking place in the cell. It was *water*. "But how does it do it?" she asked herself.

As she approached the house, she remembered her mother's promise to cook one of her favorite dishes that evening, and she wondered what "information" her mother was using to prepare dinner. The next day was going to be another wonderful learning experience, and for Alice the time couldn't fly by fast enough.

6

Water-loving and Water-fearing Molecules

Professor Holmes seemed to be in a rush when he entered the classroom and he got the class underway immediately, without the usual niceties. Noticeably absent was Kofeau with his waddling steps trailing behind his master. He had become a regular presence in the class so not seeing him surprised the students.

"In the last few lectures we discussed the central dogma of molecular biology," the professor began. "Let us now discuss a few topics that are outside the sphere of the central dogma, but are quite important to the understanding of the workings of the molecules in our cells. All these topics involve proteins. However, before doing so I want to discuss a topic that at first glance seems unrelated to proteins, and also unrelated to the central dogma, but as we go along you will realize its importance. It also has an important *moral*! A brilliant and compelling idea that prevailed for a long, long time, owing to the fact that scientists did not consider other alternatives, remaining steadfast even after the evidence as to its verity had collapsed."

Alice knew what he was driving at, and she felt quite confident that her mental faculties were 'fully charged' after the time she spent in the laboratory the previous day. She knew that the professor would discuss something involving water, but she didn't have the faintest idea exactly what he was going to talk about.

"Let us start with a very simple experiment. Suppose we have a glass with equal amounts of water and oil, say, or benzene, or whatever you like. Essentially, in this simple experiment we will make use of two liquids that do not mix, one being water and the other being some organic liquid.

"Now, add a small quantity of sugar to this two-phase system. Let us assume that we added one gram of sugar. We shall see that most of the sugar will be dissolved in the water, whereas only a small amount will dissolve in the organic liquid. We say that the sugar is *distributed* between the two liquids in favor of the water phase. Figuratively, we say that the sugar molecules 'prefer' the water phase because it is *hydrophilic* — *hydro* meaning 'water,' *philic* meaning 'loving.' Of course, there are no such feelings as love and hatred between molecules, but it is a rather convenient way of expressing the experimental fact that sugar prefers to be in water — in terms of human emotions. Thus, when we say that a solute, sugar for instance, 'loves' water, what we mean is that when distributing itself between water and oil, say, it will prefer to be in the water phase — not all of it but the majority.

"Now, let us take methane, or ethane, or propane. These are small hydrocarbons consisting of carbon and hydrogen atoms. We do the same experiment as before, and we will observe that the distribution of methane is in favor of the oil phase. In other words, the majority of methane molecules will prefer to be in the oil phase, and only a small fraction will be found in the water phase.

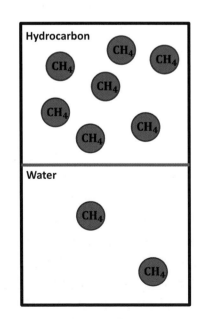

Methane, CH₄ distributed

between hydrocarbon and Water

Fig. 20 Distribution of methane between hydrocarbon and water.

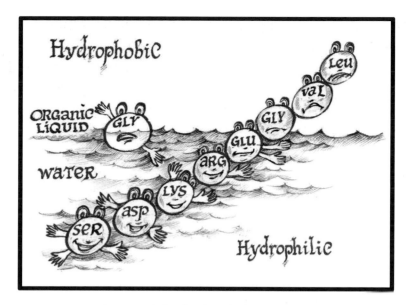

Fig. 21 The hydrophobicity scale.

We could say that methane is a 'water-hater,' or an 'oil-lover.' The term that was used for this phenomenon is *hydrophobic*, or 'water-fearing.' In molecular terms, this is not exactly the antonym of 'hydrophilic,' but the term 'hydrophobic' stuck and is widely used throughout the biochemical literature.

"So you now know that there are two kinds of molecules, termed 'hydrophilic' and 'hydrophobic.' Scientists have also suggested hydrophobicity scales. What this means is simple: take water and oil, and add a third solute. If the ratio of the concentrations of the solute is such that it favors the oil phase, it will be referred to as more hydrophobic, or less hydrophilic. On the other hand, if we find that the solute favors the aqueous environment we shall say that the solute is hydrophilic. Any questions, so far?" the professor asked.

Some of the girls were giggling, as they had a question in mind but it didn't have anything to do with hydrophobicity. The professor knew what was coming, and he asked again, directing his question to the girls. They continued giggling, until finally a boy next to them raised his hand.

"Why didn't Kofeau come to class today?" he said, grinning. "Did he submit an excuse slip?"

"Oh, Kofeau! You miss your classmate then?" replied the professor, struggling to suppress a smile. "Well, I told him over the weekend that our topics for this week would focus mainly on protein folding, and he begged to skip class as he is quite familiar with this subject. As we speak, he is in Hawaii for some brief vacation, enjoying the sun and sea and — who knows? — probably dancing the hula too! He promised me though that he would brush up on his knowledge for our forthcoming classes."

The class exploded with laughter. Kofeau dancing the hula! That really brought the house down.

"Now," said the professor, adopting the tone of a father admonishing his children, "you'd better behave or else the Dean will summon me to his office and demand to know what all this commotion is about." The laughter died down quickly and he had the undivided attention of his students once again.

"At this stage, I presume that everything is clear," he continued. "Note, however, that the scaling of hydrophobicity depends on experiment, that is, measurements of the relative concentrations of a solute in the two phases. This is purely a macroscopic definition. We aren't worried about *why* one molecule is more hydrophobic or more hydrophilic than another. This is not important for our current purposes. The fact is that some solutes are hydrophobic and some are hydrophilic. These were already well-known facts in the early 1940s. We shall also soon see that the proteins that are produced from the ribosomes must *fold* before they function. In this process, the protein behaves as if it is a string of solutes, some hydrophobic and some hydrophilic."

The professor spent the remainder of the class providing lots of examples of molecules, some more hydrophobic, others less, some more hydrophilic and others less so. There was nothing particularly exciting about the long list of molecules and the extent of their affinity towards water.

Afterwards, Alice had planned to go to the library. On the way there, she mulled over the day's class. She had understood very well the descriptive definition of a water-loving, 'hydrophilic' solute and a water-fearing, 'hydrophobic' solute. But an inner voice, or perhaps her sense of the microscopic world, told her that something was missing. Why does a molecule like sugar prefer water, whereas a hydrocarbon doesn't like being in water?

From her previous visits to the professor's laboratory, she had been given the opportunity to look at each phenomenon from a microscopic point of view. She wanted to be enlightened again this time — *see* the water-loving and water-fearing phenomena firsthand. She decided she needed to make a detour to the laboratory to get some tangible answers.

When Alice arrived at the lab, her heart sank. A note stuck to the glass door read: PROFESSOR HOLMES IN A MEETING. Disappointed, she decided to go to her original destination, the library. But halfway along the glass-lined corridor, she saw the professor heading towards her. "My lucky day," she smiled to herself.

"You caught me just in time, Alice. I have to attend another meeting but I can spare you an hour. What brings you to my lab, young lady?" he asked.

Alice explained to the professor what she had been thinking about earlier, that she wanted to take a closer look at the behavior of the different kinds of molecules.

"Well, you've come to the right place!" declared the professor as he unlocked the door and ushered her inside. He strolled purposefully over to one corner of the dimly lit laboratory, and Alice followed. "*This* might be just what you're after," he said.

Alice fixed her eyes on the large transparent booth before her, a broad smile appearing on her face as she recalled her previous journeys in the professor's 'shrinking machine.' This machine was instrumental in introducing her to the microscopic world of water last semester. It was in this machine that she first explored the 'water land.' This time, however, gone was the uneasiness she had felt in the past. With the professor's inventions, she was 100% confident that she was in safe hands — more or less, anyway.

As she stood there admiring the innocuous-looking box, it occurred to her that the professor's machine didn't even have a proper name. "From here on, I'll call it… ExploCube!" she said to herself.

Professor Holmes handed her the goggles. "Ready, Alice?"

"Ready!" she replied, and in an instant Alice was inside the ExploCube.

As soon as she put the goggles on, Alice was plunged into an astonishing new world. A little monkey was dangling precariously from a tree and waving to her with his free hand. She recognized him as he drew closer, but before she could utter a word, he began to talk.

"Hi Alice! It's nice to see you. Surprised to see me here? In Prof. H's class, I'm your classmate, but in here, I'm your guide — and always at your service!" said Kofeau. "I won't only be your guide, but I will also represent a water molecule. Call me your 'water guide'! Prof. H chose me because I have four limbs — two hands and two legs. These four limbs are a perfect match for the four arms of a molecule, don't you think?"

More than a little taken aback to see Kofeau talking, all Alice could do was nod her head in astonishment. She wasn't sure what was funnier — the normally silent, hilarious little monkey from class introducing himself as her 'water guide' or hearing Kofeau refer to the professor as 'Prof. H.' Alice knew that everything she saw was programmed by the professor. But this time, Alice thought, Professor Holmes had really outdone himself. With Kofeau to accompany her, she knew it was going to be a wild ride.

Fig. 22   Kofeau meets Alice.

"I'm sure you remember what a water molecule looks like in the gaseous phase," Kofeau continued. "It has two hydrogen atoms and one oxygen atom. Each O–H bond may be referred to as a hydrogen arm, and these are represented by my two hands. H will be referred to as a hydrogen atom of the O–H direction, and it will also represent one of my hands. When you visited the solid phase on your previous journeys in Prof. H's machine, you saw for yourself that each water molecule is in the center of a regular tetrahedron and looks as if it has four arms, two being the O–H arms, while the other two are referred to as O–L arms. In water, L represents a lone pair of electrons, but with me, L represents one of my legs.

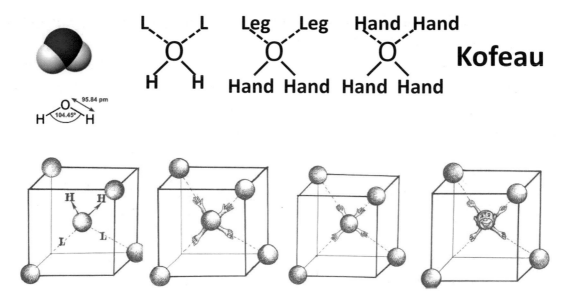

Fig. 23 Transforming a water molecule into Kofeau.

"All you have to remember is that an H arm attaches, or binds, to an L arm. Alternatively, a hand binds to a leg. In reality, the H arm and the L arm are respectively termed 'hydrogen bond donor' and 'hydrogen bond acceptor.' These are exactly the same bonds that connect the bases in the DNA." The little monkey showed her a sketch of a water molecule with four arms, two denoted H and two denoted L.

"Let's talk in simple terms," he went on. "We shall say that a hand, H, can hold and grip a leg, L. When you saw the ice the last time you travelled in Prof. H's machine, you saw one hydrogen atom flanked by two oxygen atoms. You saw the hydrogen, but you did not 'see' the L. The reason is that H is an atom, but L is just a pair of electrons. For most of the explanation of the properties of water we can even disregard the difference between L and H. We can simply view each water molecule as having four arms, which in my case means that we shall agree to view all of my four limbs simply as four arms!"

Professor Holmes' invention was turning out to be very educational, thought Alice. Her friendly little guide was as entertaining as he was interactive.

"Later," said Kofeau, "I will help you understand protein folding. But today you are still puzzled about the concepts of 'hydrophobic' and 'hydrophilic' solutes, right? Perhaps, you weren't satisfied with Prof. H's explanation about 'water-loving' and 'water-fearing' molecules — as you very well know that molecules do not possess such human emotions. You want to know how these concepts look from the 'inside' or from the microscopic point of view, am I right?"

Alice hesitated to answer the question. It was as if Kofeau's words were coming from the professor — who of course knew exactly what she was thinking. It seemed to Alice like the little monkey was just *aping* his master! "How appropriate!" she said to herself. "Kofeau the monkey aping his master!"

"All right then," the monkey proceeded with confidence, "let me show you in reality what it means to be a water-loving or water-fearing molecule."

No sooner had Kofeau said this than Alice's surroundings were transformed into an environment filled with water molecules. It was just like her

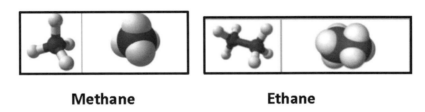

Methane   **Ethane**

Fig. 24   Methane and ethane are of hydrophobic molecules.

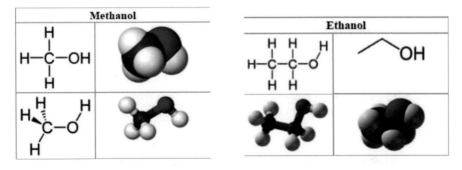

Fig. 25   Methanol and ethanol are of hydrophilic molecules.

experience in the professor's "shrinking machine" when she explored the liquid phase. All of a sudden she saw a small sphere wandering between the water molecules. At that instant all the water molecules assumed the form of little monkeys! Whenever the sphere approached a monkey water molecule, it either shrank back or it stretched out a hand to hold another monkey water molecule. Some monkey water molecules even kicked the little sphere away! It was obvious to Alice how the monkey water molecules were steering clear of the little intruder.

"You see how the water molecules like to hold hands, Alice?" said Kofeau. "In fact, a hand, H, likes to hold a leg, L, but let's ignore for the moment the distinction between a hand and a leg. You remember that Prof. H showed you two phases and a small solute inside? You can see that when that solute sphere tries to enter the water, it is kicked into the oil, or the organic, phase. Indeed, on average, very few spheres will remain in the water phase. From our point of view just think of this solute as a sphere that doesn't have any hands or legs of its own — and so isn't able to hold hands with the water molecules. We interpret this molecular phenomenon as 'water-fearing,' or hydrophobic behavior. Perhaps a better way to describe this phenomenon is to say that water molecules are hostile to molecules that don't have hands, and are therefore kicked out of the territory. Let me demonstrate what happens when a solute with hands enters the picture. You will see the difference in the behavior of the water molecules towards it."

As Kofeau said this, Alice noticed that a few new molecules had appeared, some with single arms much like the arm of a water molecule; some had two, three or more arms.

"Now you can see that whenever solutes with one or more arms enter, they are welcomed by the water molecules," Kofeau said, pointing a hairy finger. "They are 'welcomed' not in the sense that we monkeys — and humans — understand it, of course. However, the very fact that these solutes have arms enables them to hold hands with water molecules as if they become instant friends, allowing these molecules to mingle comfortably with the water molecules. On the macroscopic level, we observe that these solute molecules are highly soluble in water, so we call them 'water-loving molecules,' or

		Number of carbon atoms
Water		0
Methanol		1
Ethanol		2
Propanol		3
Butanol		4
Pentanol		5
Hexanol		6
Dodecanol		12

Fig. 26(a)   Series of alcohols.

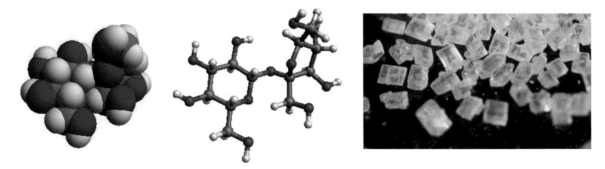

Fig. 26(b)   Sucrose.

hydrophilic molecules. In laymen's terms, perhaps a better way of describing it is a mutual love affair between solutes and water molecules. This is the way scientists like to talk about molecules, attributing to them feelings of love and hate, but in reality they do not possess such emotions.

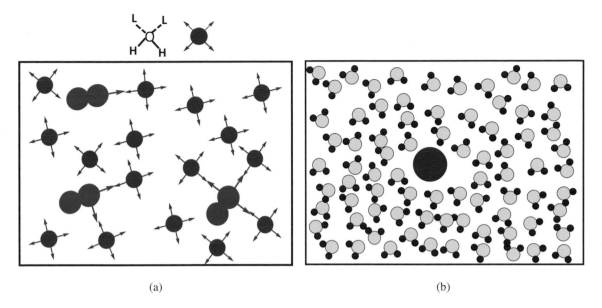

Fig. 27. (a) Water molecules with hydrophilic solutes having one, two and three arms. A water molecule represented by a sphere with four arms. (b) Water molecules with one hydrophobic solute.

"Let me summarize for you the difference between what Prof. H called hydrophobic and hydrophilic solutes. The first ones are those that are shunned by water molecules, while the second ones are very much accepted. I have only shown you one case of how solutes are welcomed, that is by their possessing one or more arms. But there are also some solutes that are electrically charged like salts. These are also very much welcome in water, but we shall not be concerned with these."

Kofeau paused for a moment and added, "You probably remember that Prof. H mentioned that sugar is a very soluble molecule. Do you know why? Well, simply because a sugar molecule has many hands, so it can attach to many water molecules simultaneously. When you learn about proteins, you will see that they have many different groups, or 'beads,' as Prof. H put it. Some have arms while others don't. These are also referred to as hydrophilic or hydrophobic. Of course, when the whole protein is immersed in water, its solubility depends on the relative amount of hydrophobic to hydrophilic groups. We shall discuss this later on. For the moment, let me discuss some simple molecules that have both hydrophilic and hydrophobic groups."

As the little monkey was talking, a new molecule entered the scene.

"This is methanol," Kofeau explained. "As you can see it has an armless sphere on one side, and an OH group on the other side. This OH group is called hydroxyl. You can look at a methanol molecule as a water molecule in which one hydrogen molecule is replaced by a sphere. Remember that water molecules have two hands, H, and two legs, L. Now what will happen if we replace one hand with a sphere — in this case the sphere is called methyl — and the whole molecule, methanol?"

"It will have one hand and two legs," Alice said confidently.

"That's right! As you would expect, such a molecule will still be very much welcomed by water. It can hold hands with its hand and two legs. Now I will show you a series of molecules. In all of these we have an OH group and a tail, a tail with no arms. What do you think will happen? Will they be welcomed by water molecules?"

Alice wasn't sure, but from the way Kofeau presented the question, she could guess the answer. "Perhaps these molecules will be welcomed only from one side," she said, "but the long, 'armless' side won't be welcomed."

"Very good, Alice!" replied Kofeau. "That is exactly what happens. Methanol, ethanol, and propanol are quite soluble in water. The effect of the three arms is enough to make these molecules welcome — and hence soluble in water. However, when the chain of 'armless' sides becomes larger, the solubility systematically decreases. In fact, even butanol has a limited solubility in water. Pentanol, hexanol, heptanol and others are less and less soluble. Think of these molecules as having 'split personalities.' One side is 'water-loving' while the other side is 'water-fearing.' I hope I have explained clearly the difference between water-loving and the water-fearing molecules."

The little monkey stopped as if pondering what to say next. "In fact, what I have told you about the various alcohols was not programmed by Prof. H," he said cryptically, after a brief pause. "I'm only saying this because it was clever of you to say that these molecules would be 'welcomed by one side, but not welcomed by the other side.' I was only expecting you to simply conclude that as the tail of the alcohol became larger, the solubility in water would become lower and lower. But what you said made me think of another important phenomena that isn't related to proteins — but something you encounter in your daily life."

Once again Kofeau paused momentarily, as if he wasn't sure whether he should continue or not. "My mission was to explain the difference between hydrophobic and hydrophilic solutes. I hope I have succeeded in that mission," he said softly. "But if you are interested, I can explain a related phenomenon that is equally important in daily life, although not directly related to proteins. However, you should keep it between the two of us."

Alice was intrigued. She knew all along that the little monkey was programmed to explain to her the things she asked the professor. She knew that all the things that Kofeau said were the professor's exact words. "Was Kofeau acting on his own?" she wondered to herself. "Was he doing something that the professor hadn't programmed? Could the little monkey think for himself?"

Kofeau's voice interrupted her thoughts. "I see that you are hesitant. Perhaps you are tired and what I have said is enough for now. If that is the case then we can call it a day. Mission accomplished," he said, confident that he had done his job.

"Oh no, Kofeau!" Alice blurted out. "I want to know what you have to say. I'd like to learn more, and I promise to keep it strictly between the two of us."

"Very well then," Kofeau replied, apparently convinced by Alice's oath of secrecy. "Now, did you wash your hands with soap before having breakfast this morning?"

"Here he goes again!" Alice said to herself. "What an absurd question! Why is he prying into my personal business?" But she was curious about where this was going, so she decided to go along with it.

"Why, yes of course, Kofeau. I washed my hands with soap this morning before having breakfast — and before each meal for that matter."

"Good! Well, have you ever wondered why you need soap to wash your hands?" Kofeau asked.

Alice was starting to get a little irritated by Kofeau's line of questioning. Everyone knows that soap *cleans*! But what on earth did soap have to do with proteins?

"I think it cleans better than just plain water, but I have no idea how it *works* exactly," she said defensively.

"Let me tell you then," Kofeau replied, smiling. "You remember, of course, the series of alcohols that I showed you. You were right in saying that when the hydrocarbon chain becomes long enough, the molecules are welcomed by water from one side, and not welcomed from the other side. This phenomenon is at the heart of the activity of various detergents, or soaps. All of these molecules have double-faceted personalities: one loves water, while the other fears water. A better way of expressing this phenomenon is the way you have said it: one part is welcomed, while the other is not. These are two ways of looking at the same phenomena. Yours, in my opinion, is the better one. This, by the way, is what prompted me to tell you about detergents."

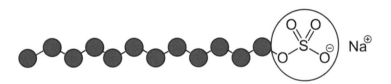

Fig. 28   Sodium dodecyl sulfate or sodium lauryl sulfate is an organic compound used in detergents.

Alice nodded, the soap concept starting to sink in.

"Now, when we have a long chain molecule, which in itself is very insoluble in water, at its edge is a hydrophilic group, such as in alcohol OH, or in carboxylic acid COCH. What happens is that these molecules are sometimes partially welcomed into the water, and the hydrocarbon part tends to be expelled from the water, into the air or into the oil phase. This is also how membranes are formed, though this is something different to the business of soap. Some of these molecules are expelled from water, yet still remain in water."

That seemed like a paradox to Alice. "How could molecules be expelled from water, and yet stay in water at the same time?" she wondered to herself.

"This phenomenon is called micellization. But don't be intimidated by this word. These molecules aggregate in such a way that all the unwelcome parts gather at the center of a sphere, while the surface of the sphere is made up of the hydrophilic groups — as you can see here," he added, showing Alice how long molecules, having a hydrophobic tail and a hydrophilic head, gathered together to form a huge sphere, the inside containing all the "tails" while the surface of the sphere consisted of all the 'heads.'

Fig. 29   Soaps.

"You see that the whole micelle is in the water. Why? Because the micelle, as a whole, exposes to water only those groups that can 'hold hands' with water molecules. At the same time all the hydrophobic groups are in the interior of the micelle, and therefore have no contact with water. So, from the point of view of the water molecule, they welcome the micelles into their neighborhood, 'unaware' that what lie hidden underneath the seemingly water-loving exterior are, in fact, all the hydrophobic groups. Effectively, they are expelled not from the water phase, but from being in contact with

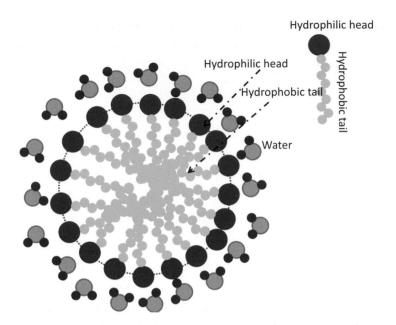

Fig. 30   A micelle.

the water molecules. Now, I hope you understand what I meant when I said that molecules are expelled from water, yet remain in water."

Alice was fascinated by everything that Kofeau had said — long chains expelled from water, yet remain in water... The picture of the micelles transported her into the ocean, and she imagined long eels, each one having a pleasant-looking head, but an awfully smelly tail. In order to *please* the water molecules and not *offend* them, the eels assembled in such a way that only their heads remained afloat in the water, keeping their 'stinky' tails away from the water. So, in a way, the eels were both in and out of the water. But then, how exactly did the soaps and detergents fit in? So engrossed was Alice in her own thoughts that Kofeau felt he had lost her.

"Alice, were you listening to me? I just explained that the detergent molecules are like long eels with a few hands on their heads and are therefore welcomed by water. On the other hand, their tails are not welcomed so they aggregate inside the micelles. Now, you can look at the whole micelle as a bubble made up of two very different parts: a hydrophilic surface that happily mingles with water, and an interior that is more like a drop of oil. But you're probably wondering what all this has to do with the cleaning action of soap…" he added, checking to see if Alice's mind had wandered again. This time, she was concentrating hard.

"Well, if you have a stain on your clothes or your hands that is not soluble in water, say a drop of oil, the micelles can absorb this oil molecule into their interior. If you rinse your hands in plain water, these oil drops will not dissolve. Oil is 'hydrophobic,' so the water won't go near it. However, if you wash your hands or clothes with detergents, some of the stain will be *absorbed* by the micelles' interior, which is like dissolving oil with another oil. The micelles as a whole carry away the oil stains with the water, thereby cleaning your hands. This phenomenon is called solubilization, meaning that micelles, with their peculiar structure, can dissolve hydrophobic molecules. Since the whole micelles are in water, the net effect is that the water is now also able to dissolve hydrophobic molecules. These molecules by themselves are not soluble in water, but they are 'solubilized' in the presence of micelles."

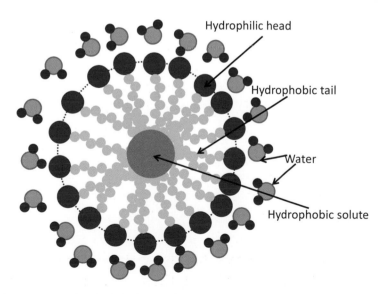

Fig. 31   Solubilization: A hydrophobic solute is accommodated in an hydrophobic environment.

Alice was dumbfounded, as she had never thought about the science behind a simple, mundane task such as washing one's hands. Realizing the importance of what Kofeau had shared with her, she was happy that she had decided to hear him out, even though she found his question about washing her hands rather strange initially. In fact, she learned more than what she had expected. In hindsight, she wondered whether what she had heard from her little friend had all been programmed by the professor — or did they come from Kofeau after all?

Her energy drained from information overload, she decided it was time to go, but not before thanking Kofeau.

"Don't forget, Alice," Kofeau told her, with a knowing look, "The last part of the excursion is just between the two of us."

For a moment Alice forgot how to go back to the real world. She recalled using a password so she could get out of the shrinking machine. Suddenly she also remembered that there was really no password and with that in mind, she immediately took off the goggles and a smiling Professor Holmes greeted her as she got out of the ExploCube.

"Oh, I didn't expect Kofeau to be back so soon from Hawaii," Alice said, chuckling, as she knew that it was merely a story that the professor had come up with earlier in the day. "What was even more surprising was to have such an informative — and vocal — guide! With Kofeau as my guide, you will see more of me in the lab."

"You are welcome here any time, Alice," replied the professor simply.

7

Alice's First Exposure to the Protein-folding Problem

Alice looked at her reflection in the mirror with satisfaction, admiring her wrist corset of miniature white roses neatly bunched in a white satin ribbon. She picked up her favorite pearl necklace from the dresser, and as she wound it around her neck the cord broke, sending the beads in all directions. Not wanting to crumple her dainty prom dress, she carefully bent down to pick up the beads, but they rolled further and further out of reach.

Suddenly, the beads started floating in the air, individually at first, and then joining together again like a chain and replicating themselves! Some of them formed sister-chains of modified beads and these modified beads then entered one side of a huge machine. On the other side was a more complicated string of very different beads of different shapes and colors. They seemed to be labeled with lower case English letters. The process continued, the machine churning out more and more beads.

Over the steady thrum of the machine, there were chattering voices, getting louder and louder... Alice suddenly opened her eyes. She realized she had been dreaming. A fascinating dream. "What message was this dream trying to convey?" she thought to herself. The pearl necklace had to symbolize the DNA, she decided. And the new string of beads symbolized the protein, the topic of today's class, perhaps?

As she sat waiting for the professor to arrive, she wondered what she would learn in class. She wanted to know more about the 'little' robots. How little is *little*, the size of a molecule? What she really wanted, of course, was to *see* them — and watch them perform their chores in the cell. The professor's voice interrupted Alice's thoughts. She was so engrossed that she hadn't even noticed that Professor Holmes had entered the room.

"In the last few lectures, we talked about the central dogma," he began. "As I mentioned earlier, this is at the heart of molecular biology. In fact, each of the steps I described consists of many steps, and it is far more complicated than the simplified scheme I showed you. But you just have to remember that information in the DNA, on the one hand, is used for replication, and, on the other hand, is used for transcription to form the RNA.

"The RNA is a go-between — between the DNA and the proteins. It transmits information contained in the DNA to the proteins. In fact, there are many different RNAs that are involved in this process, but I will not bother you with these details. You should also now realize that I use the word 'information' metaphorically. This is not the type of information that you read in a book or watch on TV. You can understand the central dogma without ever using the term 'information.' However, it somehow helps to use this term to ease understanding. I'm going to use the term 'information' again today, not in its colloquial sense, but in a metaphorical sense, in the process that is known as *protein folding*. For the benefit of those who might have forgotten, can anyone tell me where we left off in the last class?"

Alice raised her hand and said, "We stopped at the second end of the ribosomes, where a new string, which we call proteins, is formed. You also explained to us about 'water-loving' and 'water-fearing' solutes. You said that you would explain how proteins fold due to the hydrophobic effect."

"Good memory Alice," said the professor. "The existence of proteins has been known for over a hundred years. But while their chemical composition was discovered in the early 20th century, no one knew their structure. Then, when the technique of X-ray diffraction was invented, it was first applied to simple crystals, then to proteins and then to DNA. The first protein to have its structure determined was hemoglobin, that clever molecule that carries oxygen from the lungs to the cell…"

The professor winked as he emphasized the word 'clever,' and some of the students giggled. "Yes, these molecules are indeed very clever. You will see why I use this word when we discuss these clever — and efficient — molecules that transport oxygen in our bodies," he said.

"Let us go back to the fresh string of amino acids that we call 'protein.' Once X-rays were applied to protein crystals, it was established that they have well defined structures. I will show you some structures on the board, in particular that of hemoglobin. It was also concluded that in order to *function*, the protein must attain a very precise three dimensional (3D) structure. For instance, an enzyme would work properly only when it is in its proper 3D structure. Some experiments have shown that if you take an enzyme solution and heat it or add some large quantity of some chemicals, the enzymatic activity is diminished, or even destroyed. This process was named *denaturation*, meaning that the natural structure of the protein is lost. It was also known that these *denatured* proteins are 'useless' because they lose their ability to function. So, scientists concluded that in order to function properly the protein must have a definite 3D structure. A classical experiment is described in this graph. Take an enzyme, measure its activity, say enzymatic activity, raise the temperature and you will see that the activity drops sharply to zero."

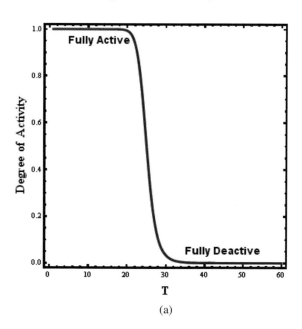

(a)

(b)

Fig. 32a   Denaturation curve.   **Fig. 32b**   Cooked and uncooked egg.

The professor pointed at the graph on the screen. "Think of any machine, or if you like a robot. What happens if it is heated?" he asked.

The students answered in a chorus: "It will melt!"

"Will the robot continue to function after being subjected to extreme heat?"

"No," came the answer, emphatically and rather confidently.

"Now suppose I soak the poor little robot in concentrated sulfuric acid," the professor continued, "Will it still be able to function?"

Again, a resounding "no" filled the classroom.

"The same thing happens to proteins when heated or put in a different solution. They will lose their *structure* and as a result they will lose their *functions*," said the professor."They will cease to function as little robots. But here is the mystery. We know that the translation of the RNA produces a specific chain of amino acids. We do not exactly know what happens to the protein that was thrown off the production line. There are too many factors that can affect the fate of the protein.

"At this point, a remarkable phenomenon was discovered by American biochemist Christian Anfinsen and his collaborators. They took isolated protein, specifically called Ribonuclease A, in controlled laboratory conditions — not in the cell but in a well-defined solution *in vitro*, which means in a glass, that is in a laboratory test tube, and not *in vivo*, which means in the real environment of the cell. They found that when the solution was gradually heated, the enzymatic activity of the protein gradually diminished until it was lost completely. This was an expected result, as the denaturation phenomenon had already been known previously. What was remarkable in the new experiment was that if one restored the original condition of the temperature and the composition of the solution, the enzymatic activity could also be restored.

"The immediate conclusion was that the process of denaturation is reversible. This means that the folding of the protein into its active or native structure can be reversed, and this occurs without any agent. This conclusion could not be drawn had the experiment been carried out on the protein in the cell, or *in vivo*, since there are too many factors that can affect the newly born protein. From this experiment, Anfinsen concluded that all the *information* necessary for the folding into the precise 3D structure is already contained in the sequence of amino acids. Subsequently, people have reached other conclusions from Anfinsen's results. First, that the 3D structure is essential for the activity of the protein. Second, that the particular composition of the solution in the experiment is essential, and — as you know — the most important component in this solution is water!

"The experiment and its conclusions opened the doors to a plethora of new questions. As I mentioned earlier, Anfinsen concluded from the result of the re-naturation that somehow the 'information' on the folding is already encoded in the sequence of amino acids. Here again we see another usage of the term 'information,' again in an abstract sense, as if the sequence 'knows' the 3D target it has to reach or acquire in order to function.

"Some people even suggested that there might be a 'code' that translates each sequence of amino acids into a 3D structure. This code, if it exists, is quite different from the code of transcription from DNA to RNA, or from RNA to protein. No one has shown that a code translating a sequence of

Fig. 33 Ribonuclease A.

amino acids to a 3D structure exists, and I doubt that it does. But this is just my personal opinion and you shouldn't take my word for it.

"The main problem that arose from the Anfinsen result is how and why proteins fold. This problem has been referred to as the 'protein-folding problem,' and was designated by the *Journal of Science* in 2005 as one of the 'unknowns of science': '*Can we predict how proteins will fold? Out of a near infinitude of possible ways to fold, protein picks one in just tens of microseconds. The same task takes 30 years of computer time.*'

"As you can see from this quotation, there are essentially two quite different problems that comprise the protein-folding problem. The first is: 'Can we predict how protein will fold?' The answer to this question depends on what one means by the word 'predict.' I can think of at least three possible definitions of this word. One, I give you a sequence of amino acids, and then you synthesize the protein and do an experiment in the lab. If it folds then you determine its 3D structure. Two, I give you a sequence of amino acids and then you do a simulated experiment on a computer. Nowadays, people refer to an 'experiment' carried out in a computer as an 'in silico' experiment. This is indeed a new concept. Anyhow, if you do this experiment and the protein folds, you get a 3D structure. Three, I give you a sequence and then you read it. You do not do an experiment but just 'predict' the structure from the sequence.

"Most biochemists will accept the second meaning of 'predict,' while a few will want to have an answer of the third kind. The second problem referred to in the quotation above has something to do with the speed of the folding process. An immense effort has been expended by many scientists to find a solution to this aspect of the protein-folding problem. One of the clearest formulations

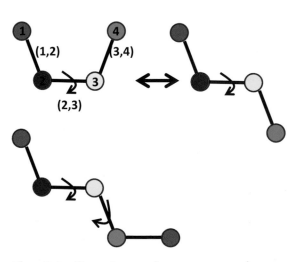

Fig. 34 Rotations about one and two bonds.

of this problem was enunciated by American molecular biologist Cyrus Levinthal in 1973. To appreciate the enormity of the problem, consider a string of beads of, say, 100 beads, which represents a relatively small protein. The beads are connected by little strings, which represent chemical bonds. The whole molecule can now rotate about each of these bonds, as indicated in this figure."

The professor proceeded to show the class a string of beads threaded through small sticks and started to rotate them about the various sticks.

"The simplest example is the case of four beads connected by three springs. Let us denote the beads by the numbers 1, 2, 3, and 4, and the springs by the pairs (1, 2), (2, 3), and (3, 4). As you can see, if I rotate the molecule about the axis (1, 2), the shape of the molecule does not change. The same is true for rotations about the axis (3, 4). You can see that all the distances between the beads and all the angles between the bonds are unchanged. However, if we rotate the molecule about the axis (2, 3), new shapes are obtained. We call these shapes *configurations*, or *conformations*. As you can see, the distances between beads 1 and 4 are altered.

"If you play with a string with six beads, you get many more configurations. Of course, for a string of 100 or 200 amino acids, there are numerous possible shapes that the protein can attain. Levinthal contemplated the following question: Supposing that the protein rotates about all its bonds at random, how long will it take to get to the correct shape of the native 3D structure?

"I have to explain what I mean by 'random change.' Imagine a drunken man who has just left a bar after downing a bottle of wine. Can you tell how long it will take him to reach his house? Of course, you cannot tell because the drunken man will be walking *randomly*, meaning he will turn unpredictably, probably not even knowing which is right or left. Clearly, if the city is very large, say like the whole of New York, it may take years and even thousands of years until he reaches his house. "Levinthal made a quick estimate as to how long it would take a protein to fold to a specific 3D structure by moving randomly in the 'configuration space.'"

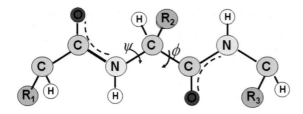

Fig. 35   A segment of a protein.

There was a buzz in the classroom, which the professor took as an indication that there was something that was not understood. Alice raised her hand.

"I do not understand what 'configurational space' is, although I understand the example of the drunken man walking in a real city, and that if he walked randomly it might take him a very long time to reach his house or any other specific point. My question is: Who is *walking* randomly, and in which space is the protein walking?"

Professor Holmes was visibly pleased with Alice's question.

"I'm glad you asked that question, Alice. It goes without saying that the protein does not 'walk' and the 'configurational space' is not the real space that we live in. What the protein does is it rotates about its chemical bonds much like the toy protein that I used to illustrate the internal motion of the molecule. Thus, 'moving' means simply changing the configuration or the shape of the protein. Scientists use the concept of configurational space to describe the collection of all possible configurations of the protein. 'Moving' in a configurational space simply means that the shape of the protein changes and 'moves' from one shape to another."

Professor Holmes took the small string of beads and started to rotate it about the 'bonds' connecting the beads.

"The important point here is that this motion is carried out randomly. There is no preferential move. If there are millions of possible shapes, then the chance that it will attain a specific shape, which we denote by N, is quite low, exactly as the case of the drunken man walking randomly in a big city. I hope that I made myself clear," said the professor, adding jokingly, "You might wish to call the randomly rotating protein a drunken protein."

Everyone laughed, the idea of viewing a protein as a drunken one lightening the mood.

"Levinthal made a rough estimate. He assumed that if the protein 'walks' at random, it would take millions or even billions of years to reach the native structure. On the other hand, we know that the protein reaches its native structure in a very short time — seconds or minutes. This apparent conflict between Levinthal's estimates and the experimental fact is known as Levinthal's Paradox.

"There was an enormous effort to solve this paradox. Before I explain the solution to this paradox, I want to mention that many phenomena in science have been explained either by 'cause' or by 'purpose.' To understand the difference between these two approaches, let us go back to the drunken man walking in the big city. Suppose the man is walking on the streets of New York and randomly turns at each intersection, either left or right, or perhaps forward and back. How long will it take for the man to reach a specific building, say the Empire State Building? Raise your hand if you know the answer."

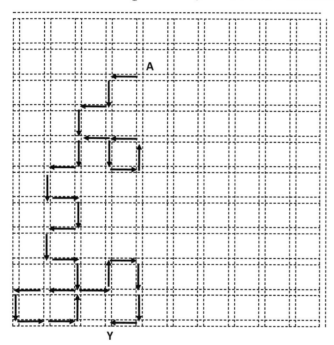

Fig. 36 A random walk in a big city.

"I think it will take a very long time. Perhaps he will never reach the Empire State Building," said John. "Of course, there is a slim chance that he will get there, but that is highly improbable."

The professor looked visibly pleased. "Thank you, John. You gave a complete and correct answer. I salute you for that."

John stood up, bowed and jokingly did a curtsey, the class laughing heartily and stomping their feet in approval. With the gesture of an orchestra conductor, Professor Holmes flapped both hands, quickly reducing the fortissimo of laughter to a mere pianissimo.

"I presume that your reaction is a sign of agreement with John's answer," the professor said with a big grin. "Now let me ask you a few more questions. What is the chance that the man will reach *any* building in New York?" The professor asked Linda to answer.

"Clearly, the man is sure to reach *any* building," she said confidently.

"Correct. You can see the difference between the two questions. If you walk randomly, the probability of hitting a *specific* place is quite low, almost zero. On the other hand, the probability of reaching one of many places is very high. We say the probability is almost one, or *there is a certainty that he will reach any building.*

"Now more questions. Again, we have the drunken man walking at random, and we observe that he reaches building No. 276 on 5th Avenue. Let us call this building Y, and let us denote A as the starting point. You can refer to my diagram on the board. I told you that the drunken man walked randomly from point A and reached building Y. Now let's bring the same guy back to point A and ask what the chance is for him to reach Y again."

The room fell silent. Everyone was thinking. Finally, John raised his hand again.

"My answer is exactly the same as before," he said.

"Excellent!" said Professor Holmes. John bowed and curtseyed again, quickly followed by zipping gesture across his mouth.

"You see, the first time we asked what the probability was of reaching *any* building, and the correct answer was 'a certainty.' Then, knowing he reached building Y, we asked again what the probability was of reaching the specific building Y, not just any building. The correct answer was that this probability is quite slim, as much as reaching the Empire State Building or any other predetermined building. Remember that the probability of randomly reaching the Empire State Building or the specific building Y is quite small, but not zero.

"Now, the next question is harder to answer. What if I told you that our drunken man started at point A, walked at random and reached point Y.

Then we put him back at point A, and again he got to Y, and so on, for a third and a fourth time. What would your conclusion be?"

Alice knew she could answer the question, and she raised her hand along with some of her classmates. Professor called her to answer.

"As you said, the probability of reaching the specific building Y by walking at random is quite slim. The probability of reaching Y for the second time from A is also very slim, presuming he walks at random. The same is true for any number of times we repeat the experiment. However, if the drunken man starts at A and reaches Y in each experiment, then I would question the validity of the assumption that the man is drunk, and that he walks at random. In other words, I would suspect that he is pretending to be drunk, and the truth is he knows his way."

"Very good, Alice. You have eloquently answered my question. What you have just said — *he knows his way* — is an example of a 'purpose' or a 'target' driven process. Let me now assure you that the man really is drunk. And let me make it even more difficult: he is also blindfolded! Nevertheless, every time he is positioned at A, he ends up at Y. What then would your conclusion be?" asked the professor, apparently testing Alice's depth of understanding.

"Well," said Alice slowly, "the person might be drunk but I would conclude that his walk is not entircly random… Something, or someone, might be pulling him towards Y."

"What could be pulling him, for example?" asked the professor.

Even before Alice could answer, Peter, who sat at the back row, shouted his answer.

"Perhaps he is blind but he has a guide dog. Or maybe it's not a dog, but rather Kofeau, who knows the target and can guide the blind man there."

"What if I told you that no one helped or guided him to reach the target, but he always managed to get there."

It was Alice's turn to speak after being interrupted by Peter.

"Perhaps the man has a keen sense of smell and some special scent emitted from target Y guides him towards the target, or perhaps the man is tied to a rope, and the rope extends all the way to Y, and someone is pulling the rope from Y."

"Very good thinking, Alice," said the professor. "But let me make it more difficult for you. As I said earlier, he really is drunk, but there is nothing pulling him towards Y. Yet most of the time when you put him at A, he ends up in Y."

"I would certainly conclude that his walking is not random," replied Alice. "If nothing is *pulling* him towards Y, then perhaps something is *pushing* him towards Y."

The professor looked very pleased.

"That answer is the best so far. In fact, it's excellent! That is almost the answer that Levinthal provided to his question. I hope you see the two different approaches to answer my question. One is assuming a 'pull' towards the target. The other is assuming that there is some 'cause' that pushes him at each intersection in some preferred direction.

"I should add, before I go back to proteins, that you are right about the fact that if nothing pulls him *towards* Y, then perhaps something is *pushing* him towards Y. But there's another slightly different possibility, and that is, he is pushed *away* from A, not necessarily at each intersection *towards* Y, but the entire pattern of pushes leads him towards Y.

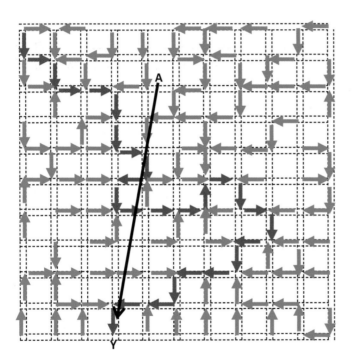

Fig. 37   A non-random walk in a big city.

"Let me explain that with a diagram. We said that nothing pulls him towards Y, but think about a very strong wind, and the possibility that the wind is in the exact direction from A to Y. Still, another possibility would be a pattern of winds as indicated by the green and blue arrows in the diagram.

The winds are always in the directions as in the figure. They are not necessarily always along the A → Y direction, but the overall result is that the drunken man walking at random is pushed along a specific pattern of directions until he reaches Y. If you like, you can think of the green arrows as the directions of the winds. The blue arrows could stand for the direction of the flow of water in a river that crosses the entire city…"

The professor paused and looked around the room.

"Shall we take a five-minute break? I see that some of you are on verge of falling asleep."

As soon as the students had settled back in their seats, the professor continued.

"Let's go back to the subject of proteins and Levinthal's question. Levinthal wondered how it was possible that the denatured protein reaches the same structure every time, i.e., starting from A, the denatured state, and reaching Y, the native structure. He estimated that if the motion were really random, it would take eons to reach Y, but in reality it takes only a few minutes. This situation was considered a paradox. But Levinthal himself did not see it like that. He recognized that the 'motion' of the protein was not random, or at least not totally random, but he did not know what it was that made the protein 'walk' on some preferred path. This was the essence of the protein-folding problem that many scientists struggled to resolve for over 50 years.

"Now, you have probably heard the story of the guy that lost his wallet while taking a walk at night. He couldn't see the sidewalk, so he went to the nearest street lamp to look for it."

When the absurdity of the story sunk in, the class broke into laughter. The professor continued.

"Well, the situation in protein folding is perhaps funnier — and more absurd. What if the wallet really fell under the street lamp but the man didn't see it? So he went looking for it in the dark — where obviously he could never have found what he was looking for! This happens quite often in science. You search, not for a wallet, but rather for a solution to the problem. You believe that it is lying somewhere, and you try desperately to find it where it cannot be found. This occurred in the case of protein folding. Many scientists searched in the wrong place.

"Some scientists thought that something *pulls* the protein towards the target, Y. Others suggested that there might be a *code* that translates from A to Y; from sequence to structure. Still others suggested that something *pushes* the protein towards the native structure, or towards building Y in our example. But no one figured out what the mysterious agent that pushes or pulls it was. A different school of thought is that evolution has created a 'code' telling the protein how to get from point A to Y, and all we need is to decipher that code.

"I should add that even in evolutionary theory, there is no target. Evolution was never presented with the problem of protein folding, therefore, it never solved that problem. Instead, during evolution some sequences were synthesized at random. Some folded, while others did not. Of those that folded, some were stable and some were not. Of those that were stable, some had evolutionary advantages, and therefore they survived. It is the result of the long journey of evolution that we now know which protein survived.

"As in the example of the drunken man, if we are not aware of any force that pushes the man along a pattern of directions, we might suspect that somehow the man 'knows' the target he has to reach. But the protein molecule does not *know* anything. If we observe in every experiment that the molecule reaches the same target, then we must conclude that there is some strong *force* that forces the protein to 'walk' along a small range of specific pathways leading to the target. Thus, what appears to us as though the molecules 'know' where they are going is only an illusion. The fact is that the protein is 'coerced' to walk in some specific preferential patterns of directions leading to the target. Look again at the pattern of the winds. Now, choose any point in the city and put the drunken man there. You will see that every time he starts walking from any initial point A, he will end up at point Y. This is of course the result of my *design* of the pattern of winds. But Nature did not *design* any pattern of forces. This pattern of forces has evolved over a very long period of time.

"Before winding up the lesson," the visibly exhausted professor told the class, "I would like to add one more thing. The clue to understanding why and how proteins fold has something to do with water. Water harbors the clues as to understanding why and how protein folds."

The professor's final statement left Alice stunned. She had followed and understood everything, but there were no hints whatsoever about the role of water in the Levinthal question, or in the Levinthal Paradox. When the professor said that *water* harbored clues, she suddenly wasn't sure she was following the professor at all. She knew that water was something of a consuming passion for Professor Holmes, but she had no idea how it entered the story of the drunken man, or the drunken protein for that matter. "Water harbors the clues…"she repeated to herself, hoping to receive a revelation. That was it! She knew it was soon time for another one-on-one tutorial in the professor's laboratory.

At that moment, Alice realized that the professor hadn't finished speaking.

"You remember the previous lecture about the 'water-loving' and 'water-fearing' solutes?" he was saying. "Now, I will tell you about the current thinking about the role of water in the process of protein folding. In 1959, Walter Kauzmann enunciated a brilliant idea. Kauzmann knew that proteins are made up of 20 different amino acids. Each amino acid can be viewed as our solute in the experiment described previously. That means we can classify all the 20 amino acids according to their hydrophobicity scale. Thus, we have amino acids that are hydrophobic and others that are hydrophilic — and some in between.

"Kauzmann noticed that when proteins fold, most of the hydrophobic amino acids are found in the interior of the folded protein, while most of the hydrophilic amino acids are exposed to the solvent. This was an important observation. It reminds us of the distinction between the two kinds of solutes. Kauzmann made the following connection between the two phenomena: the distribution of solutes — in our case amino acids — between water and oil, and the distribution of the same amino acid, but now not as a free solute, rather as one 'bead' in a long chain of amino acids. He postulated that whatever the driving force that forces hydrophobic groups to prefer oil over water, the same driving force will apply to each hydrophobic amino acid preferring the interior of the protein rather than being exposed to water. Here, the interior of the protein takes the role of the oily phase. The two experiments are shown side by side in the figure.

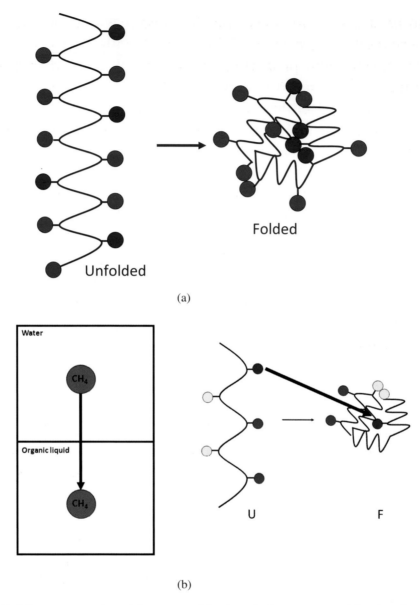

(a)

(b)

Fig. 38(a)   Folding a protein with most of hydrophobic groups (blue in the interior) and hydrophobic groups (in red, exposed to water). (**b**) Kauzmann's conjecture.

"It should be stressed, however, that Kauzmann made mental connections between the two experimental facts. There is nothing in this comparison that refers to any microscopic properties of the amino acid. In fact, Kauzmann used the concept of the hydrophobic 'bond' to refer to the tendency of the hydrophobic groups to prefer the interior of the protein. The term 'hydrophobic bond' was soon replaced by the *hydrophobic effect*, simply because there was no real bond here, and the term 'hydrophobic bond' might be misleading.

"In fact, not only was there no bond, but there was no molecular interpretation of the hydrophobic effect either. Nevertheless, the concept of hydrophobic effect gained an almost universal acceptance. Protein folding was explained by the hydrophobic effect. The association between proteins was also explained by the hydrophobic effect. We shall discuss this phenomenon in our next class. Unfortunately, the hydrophobic effect itself remained unexplained on a molecular level.

"For many years scientists were so impressed with the hydrophobic effect argument that they did not look for any alternative explanation. People imagined that the protein as a whole 'collapsed' in such a way that all, or nearly all, the hydrophobic groups were removed from water. The protein could be viewed as a micelle, but made up of a single molecule. A micelle is an aggregate of many molecules, each possessing two parts, one 'hydrophobic,' and the other 'hydrophilic.' In effect, what water does is remove the hydrophobic parts from the water while keeping the hydrophilic parts in the water.

"You remember that hydrophobic molecules prefer to be in the organic liquid, and therefore will move from water into the aqueous environment. In the same way, the hydrophobic 'beads' along the protein are in the interior of the protein, and are not exposed to water. The same is true in the process of micelle formation."

Alice was very pleased to hear the word "micelle" again. She remembered Kofeau's explanation of the effects of detergents.

"Although Kauzmann's idea was very plausible, it was never proved to be correct. The amino acids, being part of a chain of amino acids, behave quite differently from free amino acids. It is true that most of the hydrophobic amino acids are found in the interior of the protein in its native structure. However, this in itself does not provide an explanation for the folding process.

"A detailed examination of all possible solvent-induced effects shows that not only was Kauzmann's conjecture lacking, but it also revealed a rich, new repertoire of effects involving hydrophilic amino acids. For our purposes, these may be viewed as tiny arms protruding from the proteins, and reaching into the water. These arms can form bonds with water molecules.

"The moral of this story is that we should not rely only on what we see. One famous example is the theory that the earth is at the center of the universe. We see the sun rising in the east, going over our heads, and setting in the west. What could be more convincing than the conclusion that the sun rotates around the earth. In fact, for many years people believed that the sun — as well as the stars in the sky — rotated around the earth."

The professor concluded the lecture with some valuable inputs for his students to ponder on, specifically whether what is *seen* is sufficient to reach a conclusion regarding the molecular reasons behind protein folding. Those inputs filled Alice's mind as she headed for home.

8

Alice's Experience with Protein Folding

Like an excited child promised a trip to the zoo, Alice was up and about even before the cuckoo clock could do its job. At the breakfast nook was a bowl of cereals, yoghurt, a slice of melon and a mug of thick and creamy hot chocolate, all of which she devoured with gusto. She almost flew on the way to the campus.

On the way to the laboratory she thought about the drunken man walking randomly in the streets of New York. Of course, the walking man was a metaphor, but she tried to think of the analog of the water molecules that help the protein to fold in a specific pathway. Was it rain perhaps, or flooded streets that somehow flow in some pattern from street to street, sweeping the man along his path without him knowing it? If that were true, she wanted to see firsthand the role of water but on a molecular scale. She had enjoyed immensely her previous excursions into the microscopic world and she knew she would benefit from doing it again.

Professor Holmes was watering the orchids in the windowsill when Alice announced her presence at the door.

"Oh, there you are, Alice. I was expecting you to come because I know you have a good question for me. I also know that you would want to see with your own eyes how water affects the process of folding. Since you are already aware that the shrinking machine does not actually *shrink* anything, I prepared a simulated excursion for you, or perhaps I should say a guided tour into the microscopic world so you would have some idea of what goes on at the molecular level. Are you ready?"

Alice could not hide her excitement and nodded.

"Do you need some IQ boost?" the professor asked jokingly. "I'm sure you don't need it. I trust that you are confident in your abilities. Unlike your previous excursions in 'water-land,' when you had to ask *me* questions, this time your very able guide will provide you with answers. Kofeau is waiting right now."

Without hesitating, Alice entered the booth, putting on the goggles and closing the door behind her. It was eerily quiet, which made her a little nervous at first, but the thought that the professor was closely supervising her allowed her to regain her composure. A voice shattered the silence.

"Welcome to the world of protein folding, Alice!" it said grandly.

Suddenly, Alice saw Kofeau dangling precariously from a tree, waving to her with his free hand. The little monkey with big, round, smiling eyes greeted Alice warmly.

"Hi Alice! Today, not only will I be your guide, but I will also represent a water molecule. If you wish, you can also call me your water guide. I was

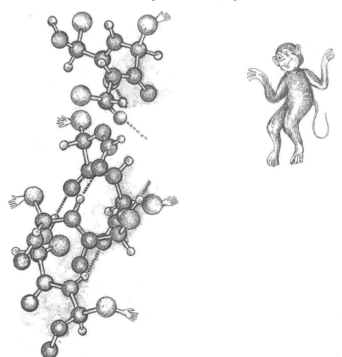

chosen to be your guide because I have four limbs, two hands and two legs. These four limbs are a perfect match for the four arms of a water molecule, don't you think?" Kofeau said proudly. "I'm sure you still recall what I explained about water molecules in the gaseous phase, as well as your visit to the solid phase of water, so there is no need for us to review those, right?"

Alice nodded.

"Now let me show you what I — and my fellow water colleagues — do once protein

Fig. 39   Kofeau with a protein.

is synthesized," Kofeau continued. "I should mention that many proteins fold with the help of some agents called chaperones or *chaperonins*." His eyes twinkled, amused by his exotic new word. "I will demonstrate to you an example of a protein that folds spontaneously. Well, not exactly *spontaneously*, but with my help. Have you heard of Anfinsen's experiment?"

"Why, yes of course, Kofeau," replied Alice. "Professor Holmes discussed his experiments in class."

"OK. So in an *in vitro* experiment conducted under well-controlled conditions in the laboratory, Anfinsen discovered that protein denatured when the temperature increased or when he added some chemicals such as urea or guanidine. Now the nice thing about his experiment was the following: In the denatured state the protein did not function. For instance, if the protein is an enzyme, one can check the enzymatic activity in the denatured state and find none, or almost nil. Now the surprising thing was when Anfinsen restored the original conditions, the protein regained its enzymatic activity, which meant that the specific three-dimensional structure of the protein was restored.

"Note that we do not know exactly in what form the protein is ejected from the production machine. We only know that after a short period of time the protein acquires its specific three-dimensional structure, so that it becomes functional or active. You can imagine that the ribosomes produce a long string of beads. As such, it can do nothing. It is 'useless.' But after a short period of time, it folds on itself, and is activated as if becoming a living robot. Perhaps you've heard about the story of Golem, which was made out of clay — but the very same inanimate object came to life. We also do not know exactly what the steps are through which the protein folds *in vivo* — whether the denaturation–renaturation of the protein *in vitro* is identical to what actually occurs in the cell. What we are striving to understand is how the protein in the *in vitro* experiment folds so rapidly into its final structure. It is clear that a major role is played by water molecules."

Kofeau added enthusiastically, "Let me show you how we do it." He exphasized the word 'we' with great pride. "Who was the 'we' he was referring to?" Alice wondered.

As he was talking to Alice, Kofeau jumped towards the protein and grabbed one of its protruding arms. Pulling the arm, he swung in a long arc and seized another arm. He paused for an instant, his familiar smile replaced by a strained expression. Alice, who had been following his every movement, was alarmed when she saw Kofeau's face change. But she quickly realized that there was no problem; it was just the enormous physical effort he has putting in to achieve what he was trying to do. After he succeeded in bringing the two arms closer together, she noticed that two other arms joined together as if they were hands clasping. Kofeau heaved a sigh of relief, as if to say he had done the job. Alice was relieved, too.

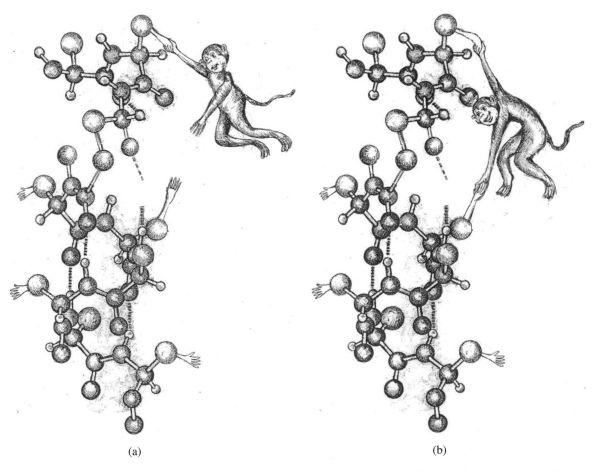

(a)   (b)

Fig. 40(a) Kofeau holding one arm of the protein. **(b)** Kofeau holding two arms of the protein.

Alice was delighted with the performance Kofeau put on for her. She suddenly understood what he was trying to accomplish. He had pulled together two 'hands' of the protein to create a new configuration that enabled a bond between a hand (H) and a leg (L) to form — and so that particular configuration was stable for a while.

"I'm sure you remember what you learned in class about hydrophobic and hydrophilic molecules," said Kofeau. "As you saw on your previous visit, the hydrophobic solutes have no 'arms,' and they are not welcomed by water molecules. On the other hand, hydrophilic, 'water-loving' solutes have one or more arms and they can hold hands with water molecules. Now, the protein can be viewed as a string of solutes, some of them having no arms, others with arms. In reality, there are many more arms in the protein than the number of amino acids.

"Thus, the traditional explanation of protein folding is that all the hydrophobic amino acids tend to be expelled from the water, and therefore the protein folds into a kind of compactly packed molecule, much like a micelle. You remember that micelles are made of many molecules, each of which have hydrophilic and hydrophobic parts.

Fig. 41 Kofeau holding a protein with two hands while a second water molecule does the same.

The protein is one giant molecule that has both hydrophobic and hydrophilic beads, so the folding is explained as a result of the tendency of all the hydrophobic beads to be expelled from water.

"What I have shown you is quite different. I showed you how the water molecules, my fellow pals, can do the folding by pulling the arms of the *hydrophilic* beads. Of course, what I did was only an illustration of what *one* water molecule can do on a very small segment in the protein, a region where two arms that belong to the protein happens to be at such a distance that a water molecule can grab one of them and bring the other closer to the first.

"This is how α-helices and β-sheets are formed. These are two periodic structures, but the result of this grabbing and pulling does not need to lead to a regular periodic structure.

"Can you imagine that a protein of say 100 amino acids has over two hundred such arms, and all these arms are surrounded by water molecules, and each of these water molecules can do the same small feat, but together their simultaneous effort can bring the apparent random configuration of the protein into its final structure, the native structure? You should also realize that at each step of the folding, the water molecules do not necessarily lead the protein towards the target. At some stages the protein may move away from the 'target.' However, the concerted effort of so many water molecules,

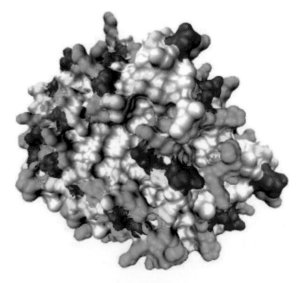

Fig. 42   An example of a protein.

simultaneously grabbing and pulling the arms on the protein eventually leads to the final 3D structure. Now, have you heard of Levinthal's Paradox?"

"That too was discussed in class by the professor," Alice said with a smile.

"OK! Well, Levinthal asked himself how a protein folds in such a short time," Kofeau continued, exhibiting his excellent mastery of the subject. "Assuming that the protein moves at random, he calculated that the folding would take billions and billions of years, and to some scientists, that seemed like a paradox. But Levinthal did not consider his absurd result to be a paradox. He simply — and correctly — concluded that his assumption of random motion was unjustified. Instead, Levinthal made an intelligent guess: the protein must proceed from the denatured state to the folded state along some preferential pathways. Unfortunately, Levinthal did not know what factor *forces* the protein to fold along a narrow range of pathways. In other words Levinthal did not know *me*, nor what I can do!"

Kofeau paused, hesitating whether to continue or not, but then said what he was going to say anyway: "I wish Levinthal had been as lucky as you are, Alice, to see me demonstrating how the protein is *forced* to fold rapidly." He chuckled but seemed satisfied with what he had said.

"You see that water can exert strong forces on the 'arms' of the hydrophilic groups, and there are so many of them, so it is easy to imagine how the water molecules would fold the protein very fast. Not only very fast, but also along a very specific pathway, eventually leading to a specific three-dimensional structure, or the native form of the protein. At that point the 'robot' can start to work."

Kofeau suddenly looked at his watch. Alice furrowed her brow as it occurred to her that a monkey wearing a watch didn't seem particularly strange to her these days.

"I have to leave now," Kofeau said, apparently in a hurry. "Do you have any questions?"

"Thank you so much, Kofeau. Your performance was magnificent! You've effectively illustrated how the water helps the protein fold. I'm very grateful. But what I do not understand is that your explanation was quite different from that of Professor Holmes."

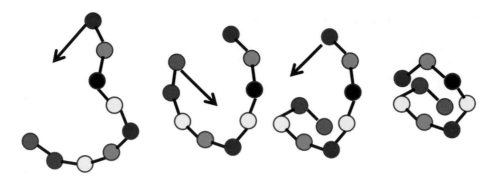

Fig. 43   Strong forces exerted on the individual groups.

"You are right, Alice. Prof. H explained the traditional or the commonly accepted view, but as you have seen, together with my fellow pals, I'm *doing* the folding!" said Kofeau, adding, "And we know better than those who see the folding from the outside."

Alice was flummoxed. But before she could ask the little monkey, he vanished, along with all the proteins. Trapped inside the pitch-black booth, she was starting to panic, reeling with confusion about what Kofeau had said, when she heard the familiar, friendly baritone of Professor Holmes' voice.

"How was your excursion today?" said the professor jovially, as the door of the booth swung open.

Alice lifted the goggles and saw the professor's reassuring, fatherly face. Suddenly snapped back to reality, she searched in vain for the words to describe the spectacle she had witnessed, struggling to make sense of the real world again — and leave the virtual one of little monkeys and giant proteins behind. That virtual reality seemed ever so real!

"It was amazing," she replied finally. "Kofeau was very helpful. I can see clearly now how the water molecules — I mean Kofeau and his pals — can help the protein to fold so quickly."

"Excellent!" Professor Holmes exclaimed. Alice gathered her thoughts, resolved to make every minute in the laboratory count.

"Everything I've seen is a demonstration on a molecular level of Anfinsen's experiment, which was carried out *in vitro*, right?" she said confidently. "So is the folding process much different *in vivo*? I guess it is much more complicated in the real cell, and that's why the protein-folding problem is still considered an unsolved mystery."

"Very good, Alice," said the professor, visibly impressed with his student's reasoning. "It's very good that you're thinking about how Anfinsen's experiment was carried out under a controlled situation, *in vitro*, in the laboratory. You might be right in your assessment that perhaps the same process, *in vivo*, might be more complicated or perhaps simpler, but we just do not know. However, the lingering mystery of the problem of protein folding is not because we don't know how protein folds *in vivo*. Even for the process carried out *in vitro*, we are still not sure how exactly the protein folds spontaneously."

"As far as I can see," said Alice, more than a little perplexed, "the folding process, as demonstrated by Kofeau, is quite simple. I don't understand why that problem is considered a big question of science."

"You are absolutely right. There is nothing mysterious about the process, Alice. The reason for the apparent lingering mystery is that many scientists spent a great deal of time and effort in searching for the answer to Levinthal's question in the 'wrong place.' You are fortunate to have taken the guided tour with Kofeau. However, you should know that for a long time people did not pay much attention to the role of water in the process of protein folding. Even after it was recognized that water is vital for the process, it was not exactly clear how water does that job."

Professor Holmes paused momentarily and then continued.

"You remember the metaphor of the drunken person walking randomly in a big city? Well, many scientists thought the protein-folding process was a target-directed process, as if there were some 'code' that commands the protein to move *towards* the target, that target being the final three-dimensional structure of the native protein. Very little effort was expended in identifying the *forces* that act on the protein. Furthermore, those who did search missed the correct cause.

"This aspect is more technical and perhaps I shall discuss it in one of our future classes. I am glad that you have grasped the main idea as to why proteins fold in a relatively short period of time to the correct three-dimensional structure. This is already an important step towards understanding one specific process. Although it is not exactly the same as the process carried out *in vivo*, we believe that the processes *in vivo* and *in vitro* are essentially the same, or at least have much in common."

With these words Professor Holmes concluded that he had said enough for the day. Alice realized she had learned more than she could have hoped for, and she thanked him profusely, once again reassured that the laboratory was a valuable extension of the classroom.

On the way home, it had occurred to Alice that there were actually two different explanations of the protein-folding problem. One emphasized the role of hydrophobic groups, which was mentioned both in class and by Kofeau. The other, as Kofeau had demonstrated, was based only on the "arms" of the hydrophilic groups, and the hydrophobic groups did not play any active role in the folding. She was not entirely sure whether Kofeau had its own view. Was he merely mouthing the words of the professor?

As she passed the field not far from her house, to her surprise she saw a horse, galloping away from its master. The animal headed towards a mound of freshly cut grass and immediately began devouring its newly discovered treasure. The master approached the horse, grabbed the reins, and led it away from its meal. With its head bowed, the horse had no choice but to follow its grumbling master. With very sad eyes, the forlorn horse stole glances at the mound of grass, hoping that his master, who kept on mumbling under his breath, would not see him.

Witnessing the whole scene, it dawned on Alice that the very concepts of 'target-based' and 'caused-based' had just unfolded before her eyes. In her mind, the target-based concept was like the horse galloping towards the mound of grass, while the cause-based concept was like the complaining master leading the reluctant horse away by the reins.

Alice suddenly became conscious of how she walked, too. Was it target-guided or cause-guided? She had no doubt that she *knew* the target she was aiming for with every step she made. People are driven by their will as they aim to make it to their targets, she thought. But proteins do not have a will, nor do they have what it takes to "know" their target. Alice smiled at the thought that even her mundane daily routine had profound connections to science.

9

How Proteins Associate to Form Large Assemblies

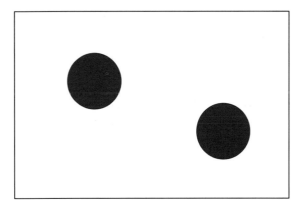

Fig. 44 The two spheres.

The lights were dimmed and everyone sat quietly. Only the vibrant colors of the two small spheres in the video illuminated the room. Apparently moving randomly, the fast-moving objects were bouncing off the walls, and from time to time colliding with each other.

"What do you see?" said the professor after running the video for a couple of minutes.

Nobody spoke. No one was really sure whether the professor was expecting an answer. On one hand, what could be seen in the clip was quite obvious: two spheres where moving around freely and randomly, and periodically colliding with each other and with the walls. On the other hand, no one really had the faintest idea what these spheres represented. Two tennis balls? Two drunken men? Perhaps two molecules in a solvent?

Professor Holmes himself broke the silence, reading everyone's minds. "Of course, you cannot guess what these spheres stand for. They could be anything you can imagine."

He pointed his clicker toward the laptop on the desk and his next video started playing. The new clips howed two balls flying freely like the first video. But when they collided, something different happened: the two spheres stuck together, and the joined pair of spheres continued moving as a single entity, never separating again.

"Can you tell me what you've just seen — ignoring what these spheres stand for?" he asked. Alice was quick to raise her hand.

"It seems that the two spheres were covered with glue, and as they approached each other, they stuck together and continued to move as one."

"Exactly, Alice," the professor replied. "You are right. Initially, the two spheres were moving independently — putting aside for the moment what they stand for — but once they got close enough to each other, there was some kind of force, which you referred to as 'glue,' that bound them together. Today, we'll discuss the nature of this 'glue.' This isn't the regular type of glue that you're all familiar with, but rather microscopic glue."

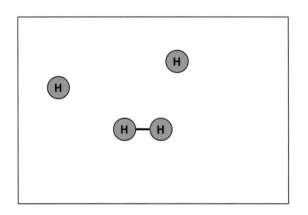

Fig. 45 Two hydrogen atoms forming a hydrogen molecule.

The word 'microscopic' immediately put a thought in Alice's head. She knew that whatever the professor covered that day — whether it was clear or not — wouldn't change her mind. She would still have questions, and they would warrant a visit to the laboratory.

The professor squeezed the button on the clicker again. The new clip showed the same two spheres as before, but now each one carried the letter 'H.'

"We now have two hydrogen atoms," the professor explained. "When they come close to each other, they form a hydrogen molecule, which we denote H_2. You can see that once they are together, they stay together, practically forever. Of course, there is no actual 'glue' that binds the two atoms together. The glue that you are familiar with exists in the macroscopic world, whereas in the microscopic world, the glue exists only metaphorically. What holds the two atoms together is a chemical bond, an extremely powerful *molecular glue*. Once a bond is formed, it never breaks down. Well, not exactly *never*. Perhaps it is better to say 'almost never.' Now let me go one step further.

"As you can see in this new clip, we have many hydrogen atoms, and each pair of colliding atoms sticks together until all the atoms are paired. We can liken it to a man and a woman who form a strong bond so that they are bound 'forever.'"

There was subdued laughter.

"Realistically speaking," said someone behind Alice, without asking for permission to speak, "not all pairs of men and women stay together forever. Some end in separation or divorce!"

"Thanks for the comment, young lady." Professor Holmes smiled at the comparison. "You are right. And in the microscopic world, some pairs divorce as well!"

He advanced to his next clip.

"Now, I will change the temperature. You can see that the molecules fly faster, and you can also see that some molecules *dissociate* into atoms. The higher the temperature, the larger the fraction of molecules that undergo dissociation. If you like, you can rephrase this to: The hotter it gets, the higher the rate of divorce!"

Laughter echoed in the classroom.

"Now, let me explain something very profound and very important that is valid only in the microscopic world. Atoms like to be free to wander around and visit all the points in the box. The higher the temperature, the larger the speed of the movement between collisions. So, we can say, figuratively, that their tendency to be free is larger. However, if a strong bond, in our case a chemical bond, binds the two atoms together, the pair of atoms loses some of its individual freedom. They — or it — can still wander about in the box, but now as a pair, and not as two separate individuals.

"At a given temperature there is a competition between two tendencies: one is the tendency to have individual freedom, and the second is the strength of the chemical bond that forces them to stay together. In technical terms, this is referred to as the competition between *entropy* and *energy*. The entropy is represented here by the tendency to be free, and the energy is represented by the strength of the bond, and hence the tendency to stay 'in touch.' As we increase the temperature, the tendency towards freedom increases, and the rate of dissociation, or if you prefer, the rate of divorce, increases."

Professor Holmes paused for a moment, in an apparent effort to choose the right words.

"Do not use the term 'divorce' to conclude that the rate of divorce in hot weather is higher than in cold weather. Don't forget that we are merely talking about divorce in a figurative sense. Technically, we say that molecules associate or dissociate — not the same thing as divorce!

"These processes are what we actually observe in any chemical reaction involving association. These could be two hydrogen atoms forming a hydrogen molecule, or two hydrogen atoms and one oxygen atom forming a water molecule, or three hydrogen atoms and one nitrogen atom forming ammonia, and so on. In all of these chemical reactions, the final state is a result of the balance between the two competing tendencies, the entropy and the energy, or in simple language, between individual freedom and the strength of the bond, or the glue. Entropy and energy are two fundamental concepts in physics and chemistry. We do not need to discuss their precise meaning here, but just think of the entropy as a measure of the tendency of individuals to wander freely and reach every corner of the box. On the other hand, the energy in our case is a measure of the strength with which the pair of individuals are attached, or if you like, glued to each other.

"Now you might ask yourselves what all this has to do with proteins, or with biology. Well, there is a profound mystery — or at least it was a mystery until very recently. Proteins wandering about in the solution would like to be free of any attachments, no strings attached. However, in reality we know that they form pairs, triplets, quadruplets and much larger complexes that live for long periods of time. Unlike the two hydrogen atoms, which succumb to the dictates of the powerful chemical bond that coerce them to stay together nearly forever, there are no such bonds that glue the two proteins together.

"Imagine two proteins in a gaseous phase. I can assure you that if we could have two protein molecules in the gaseous state, for most of the time they would enjoy their freedom as individuals. From time to time, they might collide and perhaps even stay together for a fraction of a second, but nothing like a permanent relationship. However, the same two proteins in aqueous solutions, at certain specified conditions, would form a permanent marriage and would stay together for a long time — nearly forever. Why? This has been a great mystery for a long time, no less of a mystery than that of the folding protein. For quite some time, it was known that water is essential for association, but what exactly is the factor that binds the two proteins?

"It was such a big mystery that the editors of *Science* magazine listed this as one of the 'Unknowns of Science' — right after the process of protein folding. It is of interest to quote the precise formulation of the problem as stated in *Science* in 2005: *How do proteins find their partners? Protein–protein interactions are at the heart of life. To understand how partners come together in precise orientations in seconds, researchers need to know more about the cell's biochemistry and structural organization.*

"What do you think is more of a problem, finding a partner or staying together? Well, I won't bother you with this question right now. Finding your partners involves many factors — some may be cultural, some random, and so on. The same is true, however, for the factors that keep a couple together 'forever,' whatever that means. In the case of proteins, the main question is not how the partners find each other, but what the glue is that keeps them attached together. There is another problem that we shall discuss later, and it has something to do with molecular recognition."

The professor had sensed that some of the students were getting restless, so he told them to take a short break.

"My last words before the break were *molecular recognition*," said the professor as he resumed the class. "The 'recognition' in that phrase is not the same as what you might guess, so do not waste time on this concept. We shall relegate it to a later discussion.

"For many years the self-association of proteins to form bigger units such as hemoglobin, which we shall also discuss later, was quite a mystery. Two single proteins in vacuum or in a non-aqueous solution will not associate. They are free to wander around, with no strings attached. But in a pool of water, they do associate. It looks to us as though the water serves as a glue to bind the proteins. Again, 'glue' is used here as a metaphor.

"But what factor coerces the proteins to sacrifice their freedom and form a strong, stable and long-lasting pair? You might think that there might be some chemical bond similar to the one in the hydrogen molecule, but no such strong chemical bond was found. You might think that there is some 'match-maker' that helps the pair to find each other and makes sure that they will also stay together. Or perhaps the two proteins love each other so much that they simply love to stay together voluntarily. But why do they love each other only when they are swimming in a pool of water? It may come as a surprise to you but the concepts of 'love' and 'hate' were in fact used in connection with protein binding. They were used not in the same sense that we use them in human lives, but metaphorically."

Showing another slide, the professor continued, "We see here two proteins after being synthesized and going through the process of folding. Because of their nearly spherical shape, these proteins are referred to as globular proteins. If there are enough proteins around, they will surely collide with each other. The larger the density of the proteins, the larger the frequency of collision. Thus, the main problem is not *how* the proteins find their partners, but rather what the factors are that bind the proteins together for long periods of time. This is one mystery. No one has found a strong chemical bond between the two proteins. Unlike hydrogen atoms, which are spherically symmetric, the surface of the protein is very different along different directions. The other problem here is the problem of molecular recognition, which essentially means how the proteins 'know' which surfaces should face each other in forming the stable pair, or if you like the stable couple.

"If there is no chemical bond between the two proteins, scientists suspected that the solvent — water, in particular, which is the main component in the solvent surrounding the protein — is responsible for binding. In other words, the solvent serves as the 'glue' that binds the two proteins. That conjecture is correct, but how does this glue work on a molecular level?

"As in the case of protein folding, the search for an answer to this question went astray. However, an idea captured the imagination of scientists, and that is the hydrophobic interaction. The idea is that the two protein molecules do not like to be in contact with water molecules — they 'hate' water. Therefore, by binding to each other they minimize contact with water."

"I should also tell you about an important finding that supported this conjecture. This is the story of sickle cell anemia. Hemoglobin is an important protein and is among the best known. We shall discuss its function as an oxygen carrier later. I want to tell you the story of an abnormal hemoglobin molecule referred to as hemoglobin S, or sickle cell hemoglobin. This story is very exciting since it was the first time it was speculated that a disease, called sickle cell anemia, could result from a small change in the sequence of proteins.

"More specifically, it was found that in the abnormal hemoglobin S, one amino acid, specifically glutamine is replaced by valine. The exact chemical difference is not important at this point. The important aspect of this replacement is that one hydrophilic group is replaced by a hydrophobic group. It was shown that this replacement caused excessive association of the abnormal hemoglobin, leading to a deformation of the red blood cells — erythrocytes — from roughly disk-like to a sickle-like

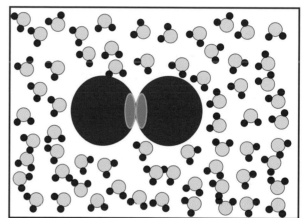

Fig. 46 Two protein molecules. The hydrophobic part of their surface (green area) is removed from the water upon association.

form. As the latter cannot flow as easily through the capillaries, the result is a painful, debilitating and often fatal disease, known as sickle cell anemia.

"I brought up this story to show you how a small change in the surface of a protein can lead to an abnormal aggregation, or if you like, an abnormal 'glue.' The factor in question that made the 'gluing' possible is the hydrophobic group in the abnormal hemoglobin. Thus, the cause of an abnormal association of hemoglobin supported the conjecture that the 'gluey' components that bind proteins are the hydrophobic groups. This makes sense because if part of the surface of the water consists of hydrophobic groups, they will be rejected by the water molecules. If the protein molecules associate in such a way that the hydrophobic parts are in the interface of the pair of molecules — you can refer to the figure on the board — the hydrophobic groups will be removed from water and will be buried in the interface between the two proteins, much like what we saw in the case of protein folding where we observed that hydrophobic groups are buried in the interior of the protein.

"This was so appealing that most textbooks and research articles adopted it to explain both protein folding and protein–protein association. Does that remind you of something?" the professor added.

No one spoke. Alice, however, had an inkling of what the professor was driving at. She recalled that during his class on the problem of protein folding, he had used the phrase "the search for an answer to this question went astray," and had gone on to explain that people held the idea that the hydrophobic groups tend to go into the interior of the protein — and that was the explanation for protein folding. Unsure what the analogy was of the interior of the protein when discussing protein–protein association, Alice decided not to raise her hand.

"I see that no one remembers the reason that has hindered scientists from finding the solution to the problem of protein folding..." said the professor slowly, and almost immediately, Alice raised her hand and began speaking.

"Yes, I do remember you saying that water-fearing amino acids tend to avoid the water environment, and therefore they bury themselves in the interior of the protein. But here we have a different situation. The water-fearing amino acids are *already* buried in the interior of the globular protein, so the water is already excluded from their surroundings. Therefore, I do not see how this hydrophobic effect can be applied to the case of association."

Hearing Alice's answer, Professor Holmes smiled broadly.

"You're right, Alice. The analogy isn't perfect. What I did not emphasize before is that in reality not all water-fearing — hydrophobic — groups find themselves buried in the protein's interior. Some are still exposed to water.

Thus, in the globular protein you can find some surface area that is composed of hydrophobic groups."

As he talked the professor showed a protein on the screen with some hydrophobic and some hydrophilic groups on its surface.

"Now, you will see that when the two globular proteins are separated, the hydrophilic groups on their surfaces are happy to be exposed to water. On the other hand, the hydrophobic groups are not happy to be in contact with water. These groups are similar to the hydrophobic groups in the unfolded protein that went into the interior of the protein. But now those hydrophobic groups that were left behind are frustrated since they are still being exposed to water. What do you think will they do?" he asked, and without waiting for an answer, continued, "Well, almost exactly the same as in the folding case. If two proteins can associate in such a way that the hydrophobic areas on the surface come together, they will repel the surrounding water molecules. In forming the pair, the analog of the interior of the protein is now the interface between the two proteins. In both cases, the hydrophobic groups, which were initially exposed to the water, are shielded from the water and feel more comfortably amidst their kin."

Professor Holmes showed two proteins approaching each other with their hydrophobic surfaces facing each other.

"As in the case of protein folding, this explanation was rather appealing. Many scientists asked themselves what could be more convincing than this explanation. In both cases, the water-fearing groups repelled the surrounding water molecules. Unfortunately, while calculations showed that this removal from the water was indeed a real phenomenon, the effect was not strong enough to explain why the two proteins would sacrifice their freedom without 'knowing' what the trade off was. In other words, when two proteins bind together, some of the hydrophobic groups are indeed in the interface between them, and therefore they achieved their aspiration to avoid any contact with water. However, it was not clear at all whether this factor was sufficiently strong to hold the two proteins together for a long time.

"It turns out that the water-loving, hydrophilic groups on the surface of the protein, rather than the hydrophobic groups, are the ones that drive the binding of the protein and 'pay' for the loss of freedom of the single protein. I shall discuss this in our next class. Meanwhile, try and think how the hydrophilic groups might do that, bearing in mind that water molecules play an active role in the association process."

The professor quickly left the room, leaving several students — including Alice — to digest the professor's words. What could the alternative explanation be?

Alice thought she knew exactly how the water molecules do the job of binding the proteins together, serving as the ultimate molecular glue between the two proteins. It struck her that "glue" was not the right metaphor. Glue was something that fills the interface between the two binding partners, but in this case she knew that water was *excluded* from the interface, and so it could not serve as glue. On the other hand, the professor did say that water molecules contributed to the stability of the protein pairs, so they must be working outside the interface region. The word "stitch" stood out in her mind. She wasn't sure how exactly they might be stitched together, but she had a strange feeling she was right.

Alice gathered her stuff together and headed straight for the laboratory. She wasn't going to let the idea escape her.

With his classes finished for the day, Professor Holmes was sitting comfortably in a chair by the window, one hand skimming through some scientific papers and another clutching a warm mug of honey lemon tea, when Alice made a sudden appearance at the door with an apologetic look on her face.

"I'm so sorry to barge in like this, professor," Alice blurted out.

"What brings you here today, Alice?" the professor replied with a smile. "Was there anything unclear about the class?"

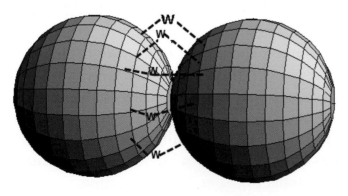

Fig. 47 Two protein molecules are 'stitched' by water molecules.

"Oh no, everything was fine," said Alice. "I just felt that I read something between the lines. What I mean is that although you did not say it, it dawned on me that perhaps you were insinuating something similar to the 'going astray' that you mentioned before in connection with the protein-folding problem... Then I felt that perhaps I was wrong, maybe I read too much into what you said and it wasn't actually there..."

Alice nearly always had several ideas running through her mind at once, and talking fast seemed to be the only way she knew how to ensure she didn't miss anything. Unfortunately, what emerged wasn't always too easy for other people to understand — particularly learned professors.

"I'm glad you noticed that there is an analogy between the problem of protein folding and protein association. In principle, you're right," the professor said, interrupting her babbling. "In fact, I was planning to prepare a new guided tour for you with our friend Kofeau."

Alice's eyes lit up, and Professor Holmes smiled.

"The analogy is in fact on two levels," he continued. "On the first level, in both cases people went astray with the so-called hydrophobic dogma. That is what I indicated in the class. People believed that the tendency of the water-fearing amino acid to avoid water was responsible for the two phenomena. The other 'level' — which I didn't talk about, but you have probably deduced from what I said about the first level — is that both of these phenomena could not be explained satisfactorily in terms of hydrophobic effects. On the other hand, in both cases the hydrophilic effects came to the rescue, turning a completely mysterious phenomenon into a clear and almost obvious one.

"Now, you'll have to excuse me, I'm afraid, Alice. I'm a little tired now, and my throat is sore. I think it's better that you come some other time. I'll arrange another guided tour with Kofeau, which should clarify this issue. I hope that he will respond favorably to my request," he said with a grin.

On her way home, Alice thought deeply about what the professor had said, secure in the knowledge that after her next visit to the laboratory, everything would become clearer.

That night, Alice's dreams were vivid kaleidoscopes of color filled with pulsating, vibrating, playful molecules. But they were no ordinary molecules, and from the different expressions on their faces Alice could sense how they were feeling.

At the beginning, they were moving together as a group, and as they approached what seemed to be a pool, those with happy faces immediately dove into it. The rest, scared stiff at the sight of the water, retreated as fast and as far away as they could. Then everything descended into chaos! Some molecules seemed to be enjoying each other's company, but some looked like they were chasing one another!

Alice woke up with a start, the bizarre images from the dream seared in her mind. She felt like she'd hardly slept. Those 'water-loving' and 'water-fearing' molecules had put on quite a show! But now she was even more confused. Professor Holmes' words from the last class echoed in her ears: *The water molecules act as glue that binds the proteins together. But this glue does not reside in between the proteins, so how do water molecules perform their gluing function?* How indeed?

She was determined to clarify matters with the professor. She was certain that paying a visit to the professor in his laboratory would make sense of her muddled views and provide answers to questions she was struggling to formulate.

10

Alice Takes a Guided Tour to the Protein Assembly Line

A smiling, relaxed Professor Holmes greeted Alice in his laboratory on the first day of the week.

"Good morning, my dear. I trust you had a wonderful weekend!" he said cheerfully. "You're fully recharged for yet another tour, yes? Are you ready?"

"Oh yes! I've been looking forward to meeting Kofeau again — and benefitting from his charm and wisdom!" she replied with a chuckle.

Alice knew the procedure well and didn't even wait for the professor's signal. She immediately grabbed the goggles and stepped into the professor's unassuming booth, the mysterious shrinking machine she'd christened *ExploCube*. She had spent the weekend rehearsing a long list of questions but suddenly realized the futility of it all when the little monkey darted out of nowhere and started talking.

"Hi Alice! We have yet another interesting tour today!" Kofeau announced excitedly. "We shall see how water molecules help in the assembly line of proteins. I know you have plenty of questions, but let's see the assembly line for ourselves first, and after that if you still have questions, I'd be very glad to help you out. I'm confident, however, that all your questions will evaporate when you see what happens!"

Alice grinned. Kofeau had taken the words right out of her mouth. That was hardly a surprise though, she thought to herself, because he was, after all, under the command of his master, Professor Holmes. On the other hand, her little guide seemed to have his own opinions about certain things. Did he perhaps have his own personality? Should she be telling Kofeau every single detail about her conversations with the professor? Alice wasn't sure.

"You're right, Kofeau, I told Professor Holmes I have many questions," she said, without saying too much, "but let's see what happens first!"

Before she could even finish her sentence, Alice's surroundings filled with a dazzling assortment of molecules of different shapes and sizes — all suspended in what she immediately recognized as water molecules. It was an eye-popping scene!

So captivated was Alice that she hadn't noticed that the little monkey was rapidly shrinking into the distance, and in a matter of seconds he had disappeared into the solution. When she suddenly realized Kofeau had vanished, she panicked, thinking she had lost her guide. Then she heard a tiny voice.

"These are globular proteins!"

The voice was coming from the far off in the distance. It was Kofeau! Suddenly, Alice recognized him, although he was so small that he almost resembled one of the little water molecules.

"This is how they look after the folding process that you observed on your previous excursion," he said. "You can also see that many hands and legs are protruding from the surface of the protein. Most of them hold hands with water molecules. You will remember that there are two kinds of 'arms': hands, H, and legs, L. The rule is that a hand always holds a leg, but for today's demonstration we can disregard the difference between a hand and a leg and just think of them as being equivalent. Now, I am going to show you what water molecules can do with their arms."

Suddenly, Kofeau jumped towards a protein and grabbed on to one of its arms. Alice couldn't help thinking that it might be a leg. He was hanging and swinging for a few seconds as he clung onto the protein's arm, much like the other molecules were doing with their hands and legs. Then when a second protein approached, he grabbed the second protein with his free hand, while continuing to hold onto the first protein. Alice thought he looked quite funny with two outstretched arms holding onto two huge proteins a hundred times bigger than he was.

Fig. 48   Kofeau binding two proteins.

"You see!" gasped Kofeau. "I'm forming a bridge between the two proteins much like the bridge that I made with the two groups on the same protein when I demonstrated the folding process. Now, I'll need to ask the help of my fellow water molecules in order to hold onto these two huge proteins for long. As you can imagine, I'll be quite exhausted in a few seconds. If I don't get help very quickly, these two molecules will fly apart when I get too tired…"

Alice chuckled to herself hearing Kofeau's reference to his "fellow" water molecules. Here he was, her guide, explaining how water functions as glue, yet at the same time he was a part of the grand scheme of science that he had been talking about!

Just as she was contemplating Kofeau's fellow molecules, Alice noticed, to her delight, that they were transforming into tiny monkeys like Kofeau! And just like him, they began jumping towards proteins with two hands outstretched, holding onto one protein with each hand. Very soon, the efforts of the little monkeys could be seen clearly. The resulting sequence looked to Alice like a series of bridges. Or perhaps like a zipper? At that moment, Alice remembered what had happened in her dream. The water molecules looked like *stitches* connecting the two proteins! Alice was enthralled. It was even more impressive than the folding process that she had previously observed.

It was fascinating to see what looked like many tiny bridges connecting the two proteins. Still, she could only see from one side, and she wondered what lay behind them. Though it clearly looked like stitches from a distance, she couldn't tell whether the monkey molecules were using their hands or their legs to form the stitches. She also realized that some used two arms while others even used three.

"As you can see, one water molecule cannot do much," Kofeau continued. "I can hold one, two, or even three arms for a short period of time, but after that the proteins will slip away. However, cooperating with my fellow water molecules makes a big difference. Together we can exert a powerful force that can hold the proteins together. Each of my fellow friends can hold two or three arms for a while and then rest."

As he said this, Kofeau relaxed his grip on the arms of the two proteins, and nothing happened. They remained stuck together.

"While I rest for a moment, other molecules can pitch in and hold the arms of the proteins. The same process repeats itself at all the other sites, so statistically the net effect is that there are always enough water molecules that do the job of bridging the two proteins. I say *statistically* because this is a dynamic process, and what you might see from afar is that the proteins are still bound — an intact sequence of stitches. In reality, if you come closer, what you will see are water molecules exchanging roles continuously. Some hold two arms, while others hold only one arm but try to grab another arm. Some get tired and go away but are instantly replaced by eager newcomers who perform the same task. So what you see from a distance is quite different from what you see up close. I hope that you can now understand the sense in which the water molecules serve as glue. It is not glue that fills up the space between two proteins, but only glue in a metaphorical sense."

As Alice watched the two proteins stitched together by around a hundred little monkeys, the scene gradually changed before her eyes. One of the monkeys began to grow in size and seemed to be coming towards her, while the other little monkeys slowly transformed into water molecules. She realized that the expanding monkey was Kofeau! It was similar to what she had witnessed when Kofeau moved further away from her as he plunged into the solution. The only difference was that initially there were only two proteins moving around in the solution, and afterwards the two proteins were bound together and seemed to be moving as a single entity.

"I'm sure you've noticed that there is no direct chemical bond that holds the proteins together," said the full-size Kofeau, his voice back to normal. "This is not like two hydrogen atoms forming a hydrogen molecule as a result of chemical bonding. It may be that a few arms of one protein hold onto a few arms of the second protein, but this is not a very significant effect. It is certainly not sufficient to hold the two proteins together for any length of time. What holds the two proteins together for a long time is not the direct *forces* between the proteins, but rather by indirect bridges that stitch together the two proteins via the many arms that protrude from their surface. The net effect is a result of the combined efforts of many water molecules.

"What you have just witnessed is quite remarkable: the combined efforts of many water molecules has a dramatic effect on the protein. This is one reason why water is so essential to life. You have seen its role in protein folding, and now you have also witnessed its role in protein–protein association. I should also add that in a way the role of water is the same in both processes, which is the formation of 'water bridges.' However, in another sense it is different.

"Do you remember when I showed you how I made a bridge between two hydrophilic groups in the process of folding? I did not mention that sometimes I got tired of stretching my arms and grabbing the two arms of the proteins. What helped me was that there were some arms of the proteins that held other arms of the same protein, allowing me to rest in between. This is called intramolecular hydrogen bonds, *intra* meaning 'within the protein,' wherein one hand of a protein holds a leg of the same protein, resulting in a stable bond.

"The situation, though, is different with the association of proteins. Here, you saw that when I got tired, I could leave and rest, and another water molecule would take my place, or an adjacent pair of arms serves as a bridge. There might be a bond between the hand of one protein and the leg of the other. However, this phenomenon is not significant in protein association. It is called intermolecular hydrogen bond, *inter* meaning in between the two proteins. In the association of proteins, 'arm grabbing' and 'maintaining the stitches' are performed by water molecules. Remember that this is a statistical effect. No single water molecule holds the two proteins for a very long time. Now, do you have any questions for me?"

Despite being a little overwhelmed, Alice felt everything was beginning to fall into place — and Kofeau was to thank for it. The little monkey's explanations were very clear and informative, and she admired him just as much as she did the professor. It was funny to think of admiring the two of them individually as if they were too distinct personalities, Alice thought. She knew full well that her admiration was for one person alone, her dear Professor Holmes.

"No questions, Kofeau. Once again, you did a wonderful job!" Alice replied with a grin, adding, "I just have to digest some of the things that I saw today."

"Great, Alice," said Kofeau, smiling broadly. "Let me just add that what you have seen today is only the first step in a process that scientists refer to as self-assembly. You saw how two proteins associate. But the process can continue to four, six, ten, hundreds and even thousands of proteins. It gets more complicated, but the 'driving force' for all these processes is always the same."

"Thank you so much for your time and effort, Kofeau. I'm sure I will see you again," Alice said cheerfully.

"Kofeau at your service, always!"

Alice removed the goggles, and the professor's kindly eyes immediately met hers. He offered an outstretched hand and helped her out of the booth.

"You look tired," she heard the professor say. "Why don't you run along now and enjoy the rest of your day. Besides, I have a faculty meeting to attend in about half an hour, so I wouldn't have time to sit down with you today. However, if you still have questions, please do not hesitate to come again."

"It's all right, Professor Holmes," Alice replied. "It was another truly fascinating tour, but I must admit I still have to process everything before I know what I need to clarify with you. Have a nice day, professor."

As she headed home, Alice recalled how muddled she had been when she arrived at the lab, and how many questions had been crowding her mind. She remembered how the concepts of hydrophobic and hydrophilic effects had been so confusing. Now she felt she understood it far better.

In the folding process, proteins have most of their hydrophobic groups in their interior. For that reason, it is only natural to think that this effect is also the driving force behind the folding process. But in the association process there were not many hydrophobic groups in the interface between the two proteins. So it is the hydrophilic groups that do the job of gluing the proteins together. Thus, although these hydrophilic groups "love" water molecules, they help to bring the two proteins together, as if to remove them from the water environment.

"That," said Alice out loud to herself, "makes perfect sense!" How wonderful to enter the professor's lab looking for so many answers — and not even having the opportunity to utter a single word — but ending up with no more questions, she thought.

Alice decided to stop by the fountain on her way home. She leaned back on one of the benches and gazed contentedly at her favorite cherubs clutching at the rush of water, taking in the lovely scent of the pine trees after the rain.

11

Self-assembly of Proteins

Alice was awakened by her mother's gentle nudging. With half-open eyes she saw her mother's smiling face looking down at her.

"You must have forgotten to set the alarm," said her mother, kissing Alice's forehead. "It's already half-past six. I'm leaving now. Your breakfast is waiting for you."

Alice made it to class just in time and was relieved to see her seat was still unoccupied. The professor arrived, loaded, quite literally, with all kinds of things: string, spheres, cubes, models and various other objects resembling toy building blocks. After carefully laying down all his paraphernalia on the desk, he started to talk.

"You remember that proteins are linear polymers," he began, picking up a string of beads. "A polymer is defined as having many units. There are 20 different units and they are represented by 20 different beads. At birth, this protein looks like a string with a random configuration."

He then turned and twisted the string he was holding between his fingers, demonstrating how flexible it was, and how many possible configurations — an almost infinite number — it can attain.

"As you can see from this string of beads, we can come up with many configurations, structures or conformations — whatever you may wish to call it. The important thing to remember is that there are many such configurations, which we refer to as *denatured states*, meaning states different from those in nature. In this form, the protein is useless. In fact, it is not even soluble in water. The first step the protein must undertake is to fold into a

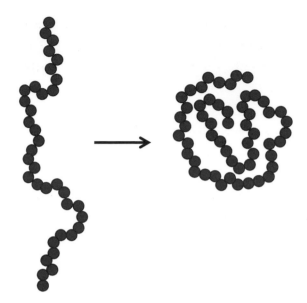

Fig. 49 A schematic process of protein folding.

precise structure. Some proteins fold spontaneously into this structure, which we call the native structure or the natured state, or the folded structure."

Professor Holmes then played around with the string, folding it until it transformed into a ball-like sphere.

"This ball represents the folded structure in the cell. Some proteins fold spontaneously while others are helped by other proteins called chaperones. For a long time the spontaneous folding process was a mystery. People wonder how the protein 'knows' how to fold into such a precise 3D structure. We have discussed the roles of water in the folding process.

"Upon reaching that final structure, some proteins can start doing their jobs. For instance, they can accelerate the rate of some chemical reactions in the cell. These proteins are called enzymes. However, many others are useless even after achieving the folded form. They must associate with other proteins, or with other molecules in the cell, to form larger and larger multi-subunit structures. We have seen how two proteins associate to form a *dimer*, that is, two units. Some proteins are functional as dimers, whereas others have to associate with more proteins to form *tetramers* (four subunits), *hexamers* (six subunits), and so forth. We have also seen how water helps proteins to associate and to maintain the bond between two proteins."

As he was explaining, the professor showed two nearly spherical proteins forming a dimer, and then he added what looked like a string of smaller beads.

"Each of these small beads represents a water molecule. Each water molecule can hold two hydrophilic groups, one on each protein, and forms a bridge between the two proteins. This is the analog of the glue that binds the protein. In reality, this is not static or permanent glue, but rather a statistical phenomenon. In other words, what you see from afar are water molecules forming a ring of stitches that connect between the two proteins. In fact, the water molecules in this ring are not permanently there, they are constantly changing partners, some leaving and others replacing them. The net effect is that, on average, there are enough water molecules to hold the two proteins together.

"If you add another solvent such as alcohol, or whatever, you gradually decrease the number of water molecules available to be 'on guard,' and to do the job of 'gluing.' Then the association between the two proteins loosens up and they will eventually separate.

"The important thing to remember is that the water molecules perform two jobs at the same time, or what we commonly refer to today as multitasking. One is firmly holding the two proteins together, which we shall call the 'gluing effect.' The second job, though no less important, is determining the *mode of association*, in other words, how the two proteins will bind, or which face of one protein will be turned toward which face of the other protein. We shall refer to this job as *molecular recognition*, as if the water molecules help the proteins to find the right faces to turn toward each other. This effect is important in many binding phenomena, such as the binding of drugs to protein, or the binding of protein to DNA. The former is very important in designing new medicines, and the second is an important aspect of gene expression, which we will talk about in our next class.

"Bearing in mind the two jobs I just mentioned, one can generalize the process of association of two proteins for a larger number of proteins. There are many examples of proteins that must first form an aggregate consisting of many proteins before they can function. I shall mention a few examples later on, but for now we should remember that if you take, say, four proteins of the same structure..."

While he talked, the professor picked up four painted balls.

"You see here four identical balls, but I have painted their surfaces with different colored patches. Clearly, there are many possibilities for the binding of even two of these, and many more possibilities for four balls. The eventual tetramer, or the four-unit aggregate, must be such that each ball must have a specific relative orientation with respect to the other three. We use the term 'molecular recognition' in a figurative sense, as if each of the balls 'knows' which part of its surface it must show to the other proteins. In reality, the proteins do not 'know' anything, in much the same way as the denatured protein does not 'know' the native structure.

"What actually happens is that the water molecules that surround the protein help to find the best, or the most stable, binding mode. Again, we say that water can 'read' the surfaces of the two proteins and bring them together in such a way that the eventual bond formed between them will be strong, or last the longest. If you like the metaphor of the stitches, you could also say that the water molecules maximize the number of stitches that connect one protein to another. The more stitches, the more stable the dimer will be, and in general this holds true for tetramers, hexamers, and so on."

Fig. 50   Lock and key.

Professor Holmes fiddled with the balls until they stuck to each other like magnets.

"I put a little magnet in each ball so that they can bind in such a way that one magnet will attract another magnet in another ball.

"Before we conclude our lesson for today, let me say that one must be careful in using arguments from evolutionary theory to explain the specific mode of binding. Today, we know that some multi-subunit proteins have highly specific functions, say the proteins in our muscles. You might think that somehow evolution has designed the protein in such a way that they will have such specific surfaces, and that these surfaces will bind in a very specific way. You might also think that evolution also faced a specific problem, and eventually it solved that problem and succeeded in the production of the required protein. Such statements are quite common, but that is not how evolution works. Evolution does not *design* anything, it does not *solve* any problem, and it does not even *face* any problem to begin with. In short, it does not strive towards a *target*.

"The fact that the proteins fold into a specific 3D structure, and the fact that the resulting protein associates in a very specific mode is not a result of a design, nor of solving any problem. Evolution doesn't work like that. Evolution means, quite simply, that things evolved. In the course of evolution, proteins were synthesized. Most of these proteins never folded. A few did fold, but only to an unstable and short-lived structure. Some that folded into a stable structure were useful; others were not. Some of those might have associated, some did not. This long sequence of events might, or might not, have led to an aggregate that might, or might not, have been useful.

"So when we encounter useful and very efficient products today, we might marvel and admire how nature or evolution achieved such a complex, efficient, useful structure. However, we should not be carried away by our admiration. These structures became what they are, not because they were *designed* to fulfill some function, but simply because, after a very, very long period of time, they were randomly — or if you like, 'blindly' — formed. Those that happened to be stable and useful have survived, and those are the ones we encounter today."

Alice was hugely impressed by the professor's very clear presentation. He was very passionate about his profession and it showed even in the smallest details, she thought to herself. Teaching was more than a profession to him, Alice decided; it was a mission. After every one of his classes, she knew she was better informed, more knowledgeable, and above all her eagerness to learn grew even more.

As she left the classroom, it dawned on Alice how her view of the world had changed fundamentally. During the first semester, she had been so focused on water. There was an added dimension and meaning to everything — whether it was rain, snow, fishermen or ice skaters. She saw water's role in all of them and was constantly trying to understand why it was so essential and vital to life. This semester had somehow redirected her focus. Instead of merely addressing how the properties of water affect life on a macroscopic level, now she was thinking in terms of the microscopic level. She knew that a glass of water could sustain human life, but more than that she could imagine how that water could trigger a chain reaction as it penetrated into each of her cells, like a colony of ants who go to work and perform simultaneous jobs in concert.

Passing the garden, with its shaped hedges and its blooming flowerbeds, she imagined water seeping into their roots and triggering a chain reaction — the end result, a riot of color.

12

Hemoglobin: Oxygen's Efficient Carrier

Fig. 51 Myoglobin.

While Alice was getting ready to go, she heard her mother talking excitedly over the phone. She could tell that it was her Aunt Sue calling to finalize plans for their vacation. Although Alice loved every minute of her laboratory explorations, she also looked forward to spending time with her mother and her relatives. With a few more weeks before the semester finished, she knew she had to maximize the time left to learn more about proteins.

Back in the classroom, Alice knew Professor Holmes felt the same way.

"We are almost at the end of the semester and I hope you have had a good glimpse at the fundamentals of molecular biology," he began. "Of course, there are many topics we have not discussed yet, and these include details of the chemicals that comprise the cell, the many chemical reactions that occur in different cells, and the many functions of proteins, both as building blocks and as robots performing a multitude of chores in each cell. In the remaining weeks of this semester, I will devote our classes to giving you a good idea of the huge repertoire of protein functions.

"The proteins that are encoded in the DNA are synthesized on the ribosomes. Once they are ejected from the ribosomes, which are largely made up of proteins, they have to fold into a specific 3D structure. We have seen how water molecules help in this highly specific process of protein folding. Once they've folded, some proteins are ready to undertake various tasks, while others must first join other proteins of the same or of a different kind, before they can tackle their jobs.

"Let's start with a brief overview of the huge repertoire of protein functions. This will give you a 'bird's eye view' of the many tasks they form in the cells. I am giving you a general idea because I don't want you to get bogged down with details, which you will learn in biochemistry anyway."

The professor dimmed the lights so his slide presentation could begin.

The first slide read: "Hemoglobin. The Efficient Carrier of Oxygen." Below the title was a schematic drawing of the hemoglobin molecule.

"How many of you have gas ranges at home and how many have electric ranges?" he asked the class.

Alice thought it strange that the professor should ask such a question. She could not see any connection between hemoglobin and kitchen stoves! Unless of course he was going to talk about the difference between a gas range that needs oxygen vis-à-vis an electric one that does not. Either way, it

Fig. 52 Hemoglobin.

turned out that the class was divided equally between gas range and electric range owners. Fascinating!

"I'm sure you are wondering why I am suddenly interested in cooking appliances," grinned the professor. "You might even suspect that the gas range needs oxygen for the flame to burn, as this slide shows. However, I will not be talking about the *burning* of gas, but rather about *transporting* it.

"It used to be that gas ranges ran on propane, which is a small hydrocarbon that produces heat when burned. Homes used to have small cylindrical tanks filled with propane. As soon as the cylinder was empty, it would be replaced by a new one. The empty tank was refilled and delivered to another house. In fact, this is still done in some countries where households can purchase individual gas cylinders for their cooking needs from various suppliers.

"Let's consider a different scenario where a tank truck serves individual homes and fills empty house cylinders directly. As soon as the tank truck is empty, it goes back to the gas depot or to a gas station to be refilled. This seems to be a more efficient way of providing gas, right?"

The professor didn't wait for an answer.

"Having said this, let's discuss the efficiency of transporting gas. The process of refilling entails a high-pressure gas flowing from an underground gas source into the truck. Clearly, the higher the underground pressure, the larger the amount of gas that will flow into the tank truck. The same is true if refilling is done in homes with cylindrical tanks: the higher the pressure difference between the tank truck and the cylinder to be filled, the greater the amount of gas that flows from the truck into the empty cylinder.

"Intuitively, you can imagine that when the total amount of gas in the truck is larger, the higher the pressure of the gas in the tank. This relationship is not necessarily a linear one, as shown in curve (a); it can be concave downwards, like (b), or an S-shape, as in (c). Look at the three curves that I've drawn on the board. Can you define the 'efficiency' of the truck in delivering the gas?"

Alice was perplexed by the professor's question. What did all this cooking gas business have to do with hemoglobin and carrying oxygen? She could not establish any connection, let alone answer the question. Only one student, Bob, raised his hand.

"I think that we can define the *efficiency* as the total amount of gas delivered divided by the cost of transportation of the gas to the different homes."

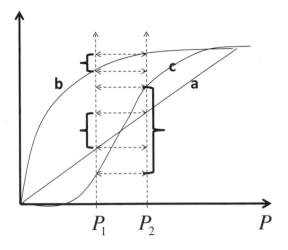

Fig. 53 The three curves and the efficiency.

"That's a very good answer, Bob," replied Professor Holmes. "But let's fine-tune this definition a little. Suppose that the pressure of the gas at the main station is fixed, and that the pressure of the nearly empty cylinder at home is also fixed. Let's also assume that the total cost incurred in the delivery — the cost of gasoline for the truck, maintenance, and some other operational costs, and so on — is also fixed. Then we can define the efficiency simply by the total amount of gas delivered. Look at the three possible curves in the figure. You will see that in (b) the efficiency is the lowest, whereas in (c) the efficiency is the highest.

"Our cells need a constant supply of oxygen, which, as everyone knows, is vital to life. Why? Without getting into details, many substances we eat are 'burned' in our cells. This 'burning' requires oxygen, the result of which is the supply of energy, which we need in order to run various processes in the body. Let us focus only on one aspect of this vital molecule, its transportation from the lungs to the cells."

Suddenly, Alice realized the connection between the transporting of cooking gas and the transporting of oxygen. She had expected to be told about the efficiency of the method of transporting oxygen — and that was precisely what she had heard!

"The pressure of oxygen in the lungs is fixed," the professor went on, "and the pressure in the cell is also fixed. Nature has found an ingenious way of transporting oxygen from the loading station, in this case the lungs, to the unloading point, the cell. The vehicle that carries the oxygen molecules in this case is the hemoglobin molecule."

Alice could guess correctly what the professor was going to say next, although she still did not quite understand the analogy between the hemoglobin molecule and the tank truck that delivers the cooking gas. One thing was for sure, though: the class was going to be interesting.

"To appreciate the efficiency of the function of hemoglobin molecules, we have to compare it with a smaller molecule, myoglobin. As you can see in the slide, myoglobin is a nearly spherical molecule, and it has one site in which a single oxygen molecule can bind. The binding site contains a specific group called *heme*, hence the name 'hemoglobin.' It is believed that in the primitive stages of evolution, myoglobin was the only carrier of oxygen. Today, we find myoglobin in the muscles of aquatic mammals such as seals and whales. Its main function is to *store* oxygen molecules to be used whenever the need arises.

"In small living creatures there is no need to transport oxygen along long routes as in the bodies of higher organisms. There is no need to be efficient. At any given time there are many myoglobin molecules; some of them carry oxygen while others do not. When the cell needs energy, it uses oxygen to burn certain chemicals. The pressure of the oxygen is lowered while the oxygen is unloaded from the myoglobin molecules. The external or the atmospheric oxygen pressure is nearly fixed. So the oxygen from the atmosphere fills the depleting myoglobin molecules. There is a typical binding curve for myoglobin, which, as you can see here, is something like curve Mb in the

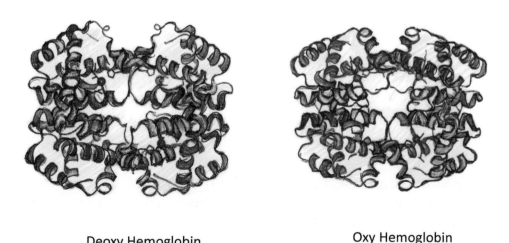

Deoxy Hemoglobin   Oxy Hemoglobin

Fig. 54 The two structures of hemoglobin: oxy and deoxy.

figure. In this figure we plot the fraction of Mb molecules that are occupied by oxygen as a function of the pressure of the oxygen in the atmosphere.

"When myoglobin is the only carrier of oxygen in small primitive cells, it can move by diffusion and reach any spot in the living organism in a relatively short period of time. In aquatic mammals, where myoglobin serves mainly as oxygen storage, the question of efficiency does not arise. However, in higher animals we need a system of transporting oxygen from the 'loading points' — the lungs — to the 'customers' — the cells — the locations of which are spread at great distances from the lungs. Therefore, we need an efficient carrier of oxygen, a carrier that will be nearly full at the loading points and nearly empty at the customer sites.

"I hope that you can see the analogy between the transporting of cooking gas and the transporting of oxygen. In the former, we transport a large quantity of gas, whereas in the latter, we transport only a few molecules at a time, but the problem of the efficiency of transportation is the same. I have said that an efficient carrier evolved from the primitive myoglobin. This is the hemoglobin, denoted Hb.

"The hemoglobin molecule is essentially a tetramer. Four units that are very similar to myoglobin are bound together. Each of the subunits can bind one oxygen molecule. The binding curve for a single subunit would look much like the curve Mb in the figure. However, when the hemoglobin molecule, in this case the tetramer molecule, works as a carrier of oxygen, its performance is a great wonder. The binding curve for the hemoglobin molecule (Hb) is shown in the figure. Oxygen pressure in the lungs, as well as in the cells, is also shown.

"If you were to supply oxygen to millions of customers at distances way too far from the loading station, which carrier would you choose, myoglobin or hemoglobin?"

"Hemoglobin," came the resounding reply. Most of the students simply guessed what the professor wanted to hear. Some looked at the curves Mb and Hb, realizing that hemoglobin would be a more efficient carrier. No one, however, understood why. The professor waited for the noise to die down and then continued.

"Let's look at these two curves. Assuming that the pressures at the loading and unloading stations are fixed, you can see that when one Hb makes a round trip from the lungs to the cell, it will deliver much more oxygen than Mb. Note that the plot is per subunit of Hb.

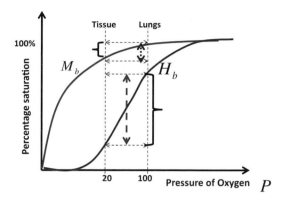

Fig. 55 Cooperative and non-cooperative curves.

"As it was shown in the early days of the discovery of Hb and Mb, the four subunits of Hb are not very much different from Mb. Yet, when they are combined into the tetramer Hb, they become much more efficient. For a long time, that phenomenon was a mystery. Think again of a truck that supplies cooking gas. Why should each truck become a more efficient carrier when it is part of a four-unit provider, rather than when it travels alone?"

The class was quiet. No one, including Alice, could imagine why a single truck became more efficient when it joined a fleet of four trucks.

"The trick that nature has 'discovered' is called *cooperativity*," the professor declared. "The theory that explains this phenomenon is now well understood, but to fully grasp the different levels of performance of Mb and Hb we need some mathematics. I shall try to explain only qualitatively how cooperativity affects the efficiency of carrying oxygen. I hope that my explanation will arouse your curiosity and encourage you to study the mathematical theory underlying the phenomenon.

"Suppose you take four Mb molecules and one Hb molecule. Call the first carrier M and the second carrier H. Suppose that at the lung the pressure is sufficiently large so that all four sites on M, and all four sites on H, are full. This means that both M and H leave the loading station with a full load of oxygen to deliver to the cells. Suppose that 100 M carriers and 100 H carriers are sent from the lungs. Altogether, all the carriers carry $100 \times 8 = 800$ oxygen molecules, 400 via the M carriers and 400 via the H carriers. When these carriers reach the cell, they unload the oxygen differently. The M carrier behaves as if it consists of four Mb molecules acting *independently*. On the other hand, the *H* carrier behaves differently.

"When the pressure of the oxygen is lower, some oxygen molecules flow from the Mb to the cell. Suppose, for concreteness, that at this pressure about 10% of the oxygen molecules is unloaded. This means that about 10 oxygen molecules from each Mb and about 40 oxygen molecules for the entire carrier, which we denoted *M* carrier. Thus, each Mb acts *independently* of the other Mb molecules, and the total unloading of oxygen molecules is simply the sum of the unloading of each of the separate units.

"The *H* carrier behaves differently. In this case the process of unloading, as well as loading, is done cooperatively, as if there is some kind of 'communication' among the subunits. After the first subunit unloads 10% of its oxygen, it sends a 'signal' to the other subunits. The second subunit releases 20% of its oxygen and sends a 'signal' to the third subunit, which releases 30%. The third subunit sends a 'signal' to the fourth subunit, which releases 40% of its oxygen. Altogether, the 100 *H* carriers will release 10 + 20 + 30 + 40 = 100 oxygen molecules, before going back to refill. Now you can see that in the round trip the 100 *M* carriers delivered 40 oxygen molecules, while the *H* carriers delivered 100 oxygen molecules. Clearly, the *H* carrier is a much more efficient carrier.

"As you noticed, I used the phrase 'send a signal,' several times but I did not explain what kind of signal is sent. Unfortunately, it is difficult to explain in a qualitative way what kind of 'communication' goes on in the Hb molecules at a molecular level. It has something to do with the molecular interactions between the subunits. The subunits in Hb do not act independently.

"What actually happens is that when an oxygen molecule binds to the first subunit, it affects its *structure*. This in turn changes the structure of all other subunits. Therefore, the second oxygen molecule reaching the Hb finds new subunits to which it binds more tightly. Then the second oxygen molecule changes the structure of the subunits on which it binds, and this affects the structures of the other subunits. Thus, the third oxygen molecule approaching the Hb binds more tightly, and the fourth oxygen molecule even more tightly. Similar events occur in the process of unloading. This is the essence of the phenomenon we call cooperativity. In a sense, you can think of Hb as four subunits that 'cooperate' among themselves to perform more efficient loading and unloading.

"Finally, I want to add a comment about evolution designing Hb to be more efficient. In fact, in the process of evolution, nature did not design the Hb molecule to behave cooperatively. Probably, nature started with randomly synthesizing proteins. Some could bind oxygen. These were found to be useful and therefore survived. Later on, when larger animals evolved, there was a need for a more efficient carrier of oxygen. It happened that a few subunits associated, and their combined behavior turned out to be more useful. Hence, there was an advantage, and hence, this survived. Thus, the Hb molecule we encounter today was not *designed* to be efficient, but it survived *because* it was efficient. We shall see that a very similar mechanism based on cooperativity also evolved in some enzymes called regulatory enzymes."

As the class ended, Alice realized that they had only just scratched the surface. Although the professor's lecture was merely meant to take a look at the big picture, Alice was tremendously excitement. She wondered whether she could 'see' how the communication between the subunits of Hb is achieved, and in what sense the subunits "cooperate" between themselves. She decided to see the professor as soon as possible.

As soon as she had returned a book at the library, Alice hurried towards the laboratory, hoping to catch the professor. She was glad to see that there was nothing written on the message board at the entrance. Professor Holmes greeted her warmly.

"You mentioned in class that the four subunits of the Hb 'communicate' with one another, and as a result oxygen transport becomes more efficient," said Alice, wasting no time. "I wonder whether I could actually *see* how this communication is achieved at a molecular level."

"An eloquently formulated question, Alice," replied the professor, evidently pleased. "Indeed, the subunits do not communicate with one another in the same sense that people or other animals do. In fact, I was thinking of demonstrating this effect to you on a molecular level, but unfortunately there is no way to 'see' this type of communication. Instead, let me explain to you what actually happens at the molecular level. I hope that after you hear this explanation you will understand what 'communication' means between molecules, and furthermore why you cannot 'see' it firsthand."

The professor took out two small cylinders and a few balls from the desk drawer.

"Suppose each of these cylinders is one subunit of Hb. I will demonstrate the effect of 'communication' with only two cylinders, but the argument can be easily extended to four or any other number of subunits. These little balls represent oxygen molecules. A full load of the Hb in this model means that there are two balls — the oxygen molecules — occupying the two sites on the cylinders — the subunits.

"Clearly, the balls do not communicate with each other. However, suppose that the two balls attract each other, like two magnets, or two charges of opposite sign. In addition we assume that the cylinder attracts the balls. Now, let us start with 'empty' hemoglobin, I mean two empty cylinders in an environment of oxygen gas. The first oxygen that arrives at the Hb is attracted by the Hb, or by the cylinder in this model. The second oxygen molecule will be attracted *both* by the cylinder and the first oxygen. Therefore, the net attraction of the second ball to the cylinder will be larger than the first.

"Of course, you cannot 'see' the force of attraction, but we can see the effect of these forces. Think of the process of unloading the fully loaded Hb. The first oxygen has to overcome the attraction exerted by both the cylinder and the other oxygen, so it will be hard for it to be unloaded. On the other hand, the second oxygen can be unloaded much more easily, because it is attracted only by the cylinder, and not by any other oxygen molecules.

"You can see that when attractive forces exist between the oxygen molecules occupying different subunits, there is a difference in the ease of loading, or the unloading of the first and the second oxygen molecule. This effect is what we call cooperativity. It looks to us as if the two oxygen molecules communicate with each other. An oxygen molecule seems to 'know' whether it is the 'first' or the 'second' to be bound to the Hb.

"In a real Hb molecule the cooperativity is much more profound and more mysterious. The reason is that the actual distances between any pair of oxygen molecules is quite large. Therefore, there are practically no attractive forces

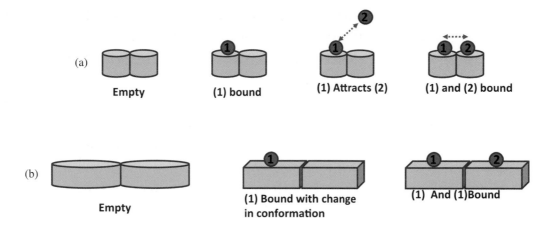

Fig. 56   (a) Binding with direct interaction between oxygen molecules. (b) Binding with conformational change.

between two or more oxygen molecules. In this case, the 'means of communication' is very different from the one that I explained to you previously."

Professor Holmes rifled through the drawer and produced a few blocks of wood.

"The communication in the Hb is more sophisticated. For a long time it was not understood. I hinted at this in our class, and I will demonstrate the effect with this model. Suppose that we start again with two empty cylinders. Just as before, oxygen molecules approach these cylinders. Again, the first oxygen is attracted to the cylinder. However, once the oxygen is attached to the cylinder, it changes its shape into a rectangular block. This rectangular block changes the second cylinder into another rectangular block.

"When the second oxygen approaches, it does not 'feel' any attraction originating from the first oxygen, simply because it is very far from it. Nevertheless, the second oxygen now binds to a rectangular block rather than to a cylinder. This means that the binding force of the oxygen is different, depending on whether it is the first or the second to approach the Hb. We can say that the second oxygen 'knows' that the other site is occupied by the oxygen molecule. Of course, the oxygen molecules do not really know anything. It only seems to us as if they 'sense' the situation at the other sites. You can see that this means of 'communication' between the oxygen molecules is very different from the one based on direct interactions between the oxygen molecules. It is often said that the second means of communication is by *indirect communication.*"

The professor paused for a moment and then concluded, "I hope you got a glimpse of what we mean by 'communication' and 'cooperation' between molecules. This is as far as I can go in explaining this effect in a qualitative manner. I hope that what I have said will inspire your curiosity to learn the mathematical theory of cooperativity. That, I'm afraid, is the only way of penetrating into the complexity of the process of the binding of oxygen by Hb."

Alice was deeply impressed by the professor's willingness to explain difficult topics by means of simple models. She thanked him and left the laboratory with a pledge to study the mathematics behind cooperativity.

13

Enzymes: Accelerators of Chemical Reactions

As Alice sat and waited in the classroom for the professor, she thought about the subjects she would be taking next semester. She would have to decide based on her progress in her current class but she mulled over consulting the professor on some related matters regarding next semester's subjects.

Some of Alice's classmates, talking excitedly about their vacation plans with their backs to the board, did not see the professor enter the room. He tapped his pen on the table to get their attention.

"The semester is ending soon and we don't have the luxury of time. So let's not waste time," he began, the room suddenly quiet. "More than a century ago, before the structure of any protein was known, the catalytic effects of certain proteins were discovered and studied in great detail.

"Basically, it was found that a certain chemical reaction, say A and B forms a new compound: $A + B \rightarrow C$ was found to proceed at a very low rate. 'Low' rate simply means that if you mix A and B in a test tube *in vitro*, and measure the rate, or the amount of the product produced per unit of time, you find that it is very low. On the other hand, it was known that the same reaction, in the cell, or *in vivo*, occurs much faster, sometimes more than a million times faster than *in vitro*. There must be some agent that 'helps' to accelerate this reaction, an agent that is present in the cell, but not in the test tube.

"To cut a long story short, scientists systematically added various components extracted from the cell and added them, one at a time, to the test tube. It was very tedious work, but it eventually paid off. They found that when some substances were added to the test tube, the studied reaction was significantly accelerated. These substances were named enzymes. Later, it was found that most of these substances were proteins, their structures were determined, and their working mechanisms were elucidated."

Professor Holmes paused for a few seconds, thinking carefully how he should explain a difficult issue with a simple analogy.

"Suppose you are at a point A,"he said, pointing at the board, "and you need to get to point B. Between A and B there is a hill that you have to climb in order to get to B. Suppose also that you have to cross from A to B several times, or that there are many of you who must cross from A to B on a daily basis. How can you *accelerate* this process? In other words, can you suggest how to get from A to B faster?"

A few students raised their hands. Alice thought she would simply run over the hill, but it seemed like too simplistic an approach to solving the problem and certainly not what the professor was expecting to hear.

"I'd drill a tunnel through the hill," said Jerry, the class clown, and a few girls giggled.

"That is a possibility," said the professor. "Any more suggestions?"

Linda said, "Perhaps we could chop off the hilltop, thereby shortening the distance between A and B."

"OK. Any more educated guesses?" the professor pressed on.

Since no one else seemed to have an answer, Alice raised her hand and said, "I would run faster over the hill."

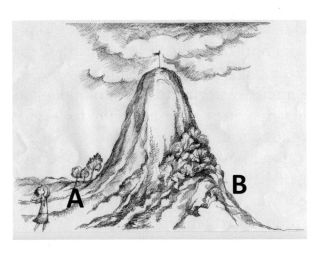

Fig. 57   Alice faces a hill.

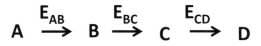

Fig. 58   Sequence of reactions.

"All of your suggestions are plausible. Alice's suggestion is equivalent to raising the temperature of the system. However, what enzymes actually do is tantamount to lowering the height of the hill. Of course, in the chemical reaction there are no hills and there is nothing to chop or drill. This is only a metaphor. The 'hill' that the chemicals *A* and *B* must cross is called the *activation energy*, and what the enzymes effectively do is to lower the activation energy. It's equivalent to Linda's suggestion of chopping off the hilltop.

"In each of our cells, there are thousands of chemical reactions going on at any given time. Many of these are catalyzed by enzymes, and these enzymes are highly *specific*, meaning that each enzyme catalyzes one specific reaction. You can also imagine that there are many enzymes that are at work simultaneously to accelerate the various chemical reactions.

"Today, the structures of many enzymes are known, and scientists have a fairly good idea as to how the enzymes do their jobs. One of the most common mechanisms is binding the reactants *A* and *B* on specific sites on the enzyme's surface, thereby enabling *A* and *B* to form the new compound *C*, which is then released from the binding site. Here is a schematic description of the way an enzyme can work.

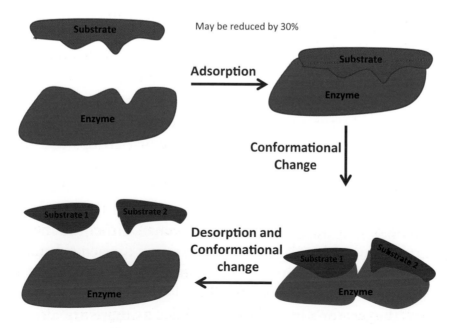

Fig. 59   Stages in enzymatic activity.

"The important point to remember is that the *specificity* of the activity of the enzyme is a result of the specificity of the binding sites to the reactants *A* and *B*. In most cases, the chemicals *A* and *B* will fit into the site much like a key fitting a keyhole. This is known as the 'lock and key' metaphor. Thus, a given enzyme will bind only with *specific* reactants, and therefore its catalytic effect is specific to the chemical reaction between *A* and *B*.

"As you can imagine, there are many different enzymes helping and accelerating many chemical reactions. There are also enzymes whose extent of activities can be changed. These are called *regulatory enzymes*. Has anyone done a blood test recently?"

The question surprised Alice. She couldn't see the relevance of a blood test to enzymes. It had been some time since she had done a blood test, but she raised her hand anyway. She was eager to understand the connection.

"I was sick a year ago and I had to do a blood test," she said. "The doctor wanted to find out what was causing my high fever."

"Have you seen the results of the test? Do you recall what it looked like?" the professor asked.

"Yes. As far as I can remember, there was a list of substances and two columns of numbers. I believe that one of the columns showed the minimum and maximum ranges for the substances, while the other column gave the actual concentration of the substances present in my blood."

"Do you also remember what the doctor said when he gave you the results?"

"I remember the doctor saying that the level of a certain substance, the name of which escapes me now, was too low, or perhaps too high, and he prescribed my medication accordingly."

"Thank you, Alice. It doesn't matter that you don't remember what the specific substance was. What matters is that you noticed that in the actual substance count, one of the substances was out of range, either too high or too low. In the past, doctors would look at the numbers next to each substance and they could tell which number was out of range simply because they memorized the 'normal' range for each of the substances while they were in medical school.

"Nowadays, the doctor's life is made simpler, as there are already two columns of numbers, just as Alice observed in her test results. One column reflects the actual count of the substances present in the blood, and the other reflects what is called the 'normal range' reference chart or numbers."

The professor pulled a sheet of paper from his briefcase, snapped on the overhead projector and placed the sheet on the glass. The following table was printed on the page:

	Result	Units	Reference values	Remarks
GLUCOSE	82.84	MG/DL	(65.00 – 100.00)	(.. . .*. . . .)
ALT. (GPT)	41.34	MG/DL	(5.00 – 40.00)	(. . . . .)*
BUN	10.23	MG/DL	(4.60 – 21.00)	(. .*. . . . . .)

"If the asterisk falls within the brackets," the professor continued, "it means the amount of substance is within the normal range. If the asterisk falls somewhere outside the left bracket, it indicates that the amount is too low, and if the asterisk falls somewhere outside the right bracket, it means the amount is too high. Now for the more difficult problem: Have you ever wondered how the body can maintain a given substance within a narrow range of concentrations — a range that we consider to be normal?

"This is not a trivial question. The answer lies in the extraordinary working methods of some enzymes, referred to as regulatory enzymes. The most surprising aspect of the mechanism of these regulatory enzymes is that it is very similar to the mechanism of cooperativity in the oxygen transport by hemoglobin.

"Let me explain the main idea underlying the workings of regulatory enzymes. Suppose that we have a series of chemical reactions that we write symbolically as $A \rightarrow B \rightarrow C \rightarrow D$. Compound A is converted to B, and is catalyzed by an enzyme denoted E_{AB}. Compound B is converted to C, catalyzed by E_{BC}. And compound C is converted to the final compound D, catalyzed by enzyme E_{CD}. Suppose that there is an unlimited amount of A and the sequence of reactions starts. What will happen?"

Everyone raised their hands, apparently without giving the question much thought.

"I see that you all know the answer. Indeed, the answer is quite simple. If there is an unlimited supply of A, then there will be an unlimited production of D, right?"

"Yes!" came the answer, loud and enthusiastically.

"OK. I will dampen your spirits a little," said the professor, grinning. "This time I have a difficult question for you. Suppose that the body can operate properly only when D is within a narrow range of concentrations. How can it do so if we still have an unlimited supply of A?"

Without waiting for answers, the professor continued.

"The answer to this question eluded scientists for a long time. However, today there are several reactions known to be regulated by regulatory enzymes. Here's how it works. The first reaction, $A \rightarrow B$, is catalyzed by the enzyme E_{AB}. As we have seen in the previous illustration, the enzyme works by binding A to a site, which is called the active site. Then the conversion from A to B occurs, and compound B is desorbed from the enzyme. Next, compound B produces C, and C produces D, in turn. Now suppose that the released compound D can also bind to the enzyme E_{AB}, not on the active site E_{AB}, but on a different site, which we call a regulatory site. The fact that D binds to a different site is the reason why such enzymes are called allosteric enzymes — *allo* means 'different' or 'other', and *steric* means 'solid'.

"The mechanism of regulating enzymes is quite similar to the mechanism of transporting oxygen efficiently by hemoglobin. This mechanism is called cooperativity. In hemoglobin, an oxygen molecule bound to one site can 'communicate' with another oxygen molecule on another site. This is a very remarkable effect since the distances between any pair of binding sites is quite large compared with the molecular diameter of the oxygen. The communication is achieved not by direct interaction between the two oxygen molecules, which is negligible at these distances, but by modifying the configuration of the protein. This is schematically shown in the following illustration.

"Initially, the subunits have one configuration, say square in the illustration. When oxygen molecules bind to one subunit, a configurational change is induced in that subunit. This subunit then induces a change in the neighboring subunits, thereby altering their affinity for oxygen. When another oxygen molecule approaches the empty subunit, it 'sees' a different subunit than the first oxygen. We say that the second oxygen 'knows' that another site is already occupied."

Alice was very happy that she had the opportunity to see this behavior in Professor Holmes' laboratory. She also knew what the professor meant when he said that the oxygen molecules 'know' whether another site is empty or occupied.

"The same principle applies to regulatory enzymes. Suppose that an active site can bind a reactant only when its shape is cubic. When a regulatory molecule is bound to a different site, it induces configurational change in the molecules, altering the shape of the active site. An extreme example is shown in this illustration. Regulator R, when bound to the enzyme, changes the active site in such a way that it can no longer bind to the compound A. Thus, the regulator effectively deactivates the binding site, and hence the entire activity of the enzyme is switched off.

"Suppose that the regulator molecule is D. In this case, you can understand how the concentration of the product D of a chain reaction can be regulated. When the concentration of D is high, it binds to the enzyme E_{AB}, and deactivates it. Hence, the whole chain of reaction stops producing D, and therefore its concentration will drop. If the concentration of D drops too much then it will desorb from E_{AB}, and the activity of the enzymes is recovered.

"One of the most carefully studied regulatory enzymes is called ATCase. It consists of six subunits, three with active sites and three with regulatory sites. The switching on and off of the enzyme is similar to the binding curve of oxygen to hemoglobin.

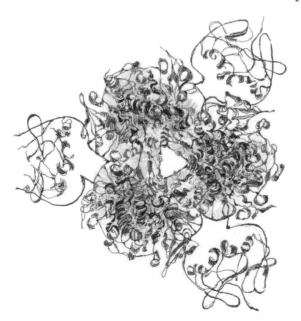

Fig. 60 A regulatory enzyme: aspartate transcarbamoylase.

"Not all the regulations are achieved by allosteric enzymes. For instance, hormones send messages when certain compound levels are too high or too low. A well-studied case is insulin and glucagon. When the sugar level is too high, insulin sends a message to lower its concentration. When it is too low, the reverse is done, and glucagon sends the corresponding message. Insulin deficiency leads to diabetes. The concentration of sugar can shoot up dramatically, out of control."

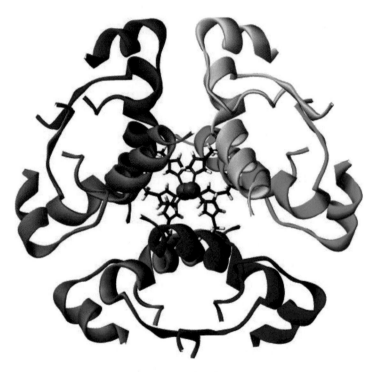

Fig. 61   Insulin hexamer.

Alice was suddenly reminded of one of her relatives suffering from the diabetes. Learning about it now gave her a new appreciation for the disease and the complex mechanisms involved.

14

Molecular Recognition

"Today, we shall discuss molecular recognition," declared Professor Holmes. "I trust that everyone in the class knows what 'molecular' and 'recognition' mean. However, when put together 'molecular recognition' is quite different from what we obtain by combining the two concepts.

"Molecules do not recognize other molecules in the sense that people recognize each other. Human recognition is a complex process that occurs in the brain. You observe the shape or color of a face or an object, listen to his or her voice, or to the sound the object makes, and you might conclude that you have identified the person or the object. Likewise, when you meet a person in the street, you receive some signals from that person such as his or her shape, sound or smell, and you can tell whether that person is familiar or not. In order to reach such conclusions, the brain processes all the 'information' it gets from the senses, compares it with information stored in the memory, and makes a decision — recognized or not recognized.

"Nothing like this goes on in the 'brain' of a molecule when it makes the decision we call 'molecular recognition.' Nevertheless, the concept of molecular recognition is extremely useful, and scientists use it in connection with almost any process of binding two macromolecules, or binding a small molecule to protein or to DNA.

"Basically, what we mean by molecular recognition is very simple. Suppose the surface of the protein looks like this," said the professor, showing his first slide. "A small molecule, say, a drug approaches the protein. It can hit the protein at many points on its surface. What will happen when it hits the part of the surface that exactly fits the shape of the small molecule?"

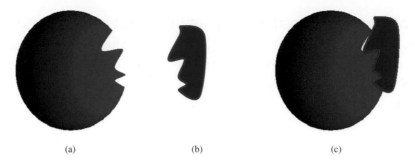

(a)   (b)   (c)

Fig. 62   Molecular recognition; (**a**) a protein, (**b**) A drug, and (**c**) the drug fits on the surface of the drug.

The professor didn't wait for an answer.

"Clearly, the small molecule will bind more tightly to the site in which it fits best. Similarly, there are many ways in which two proteins can bind, but there is one binding mode that is preferable, as shown in this schematic illustration. Clearly, the molecules do not 'recognize' each other in the sense that human beings do. However, the molecules bind at a specific location *as if* they recognize the best, or the most suitable, location. This model of recognition is known as the 'lock and key' model. Now, if you have a key, can it open any lock?"

"No," everyone said loudly.

"And what about many keys and one lock? Which one will open the lock? Of course, you all know the answer. Only the key that fits the lock will open it. Similarly, only the molecules that fit as perfectly as possible will bind. The reason is that when the two surfaces fit, there is maximum interaction

Fig. 63   Binding according the lock and key.

between the two surfaces, and maximum interaction means a more stable bond.

"As I mentioned earlier, there are actually two phenomena that require explanation. One, what factor makes the two proteins bind and stay together? Two, what determines the specific mode of binding? For a long time the metaphor of the 'lock' and 'key' was used to explain the binding of small molecules, like drugs to proteins, or proteins to proteins, or even proteins to DNA.

"We have seen how a small molecule fits tightly to a specific location on the surface of the protein. Similarly, two proteins can bind by fitting the shape of one surface — the key — to the shape of the second surface — the lock. However, when water acts as the glue, the story is quite different. Look at the following two proteins. You see here two proteins that are nearly spherical. The arrows protruding from a surface represents the 'hands' or the 'arms' of the hydrophilic group, which are abundant on the surface of the protein. Can you tell how these two proteins will bind?"

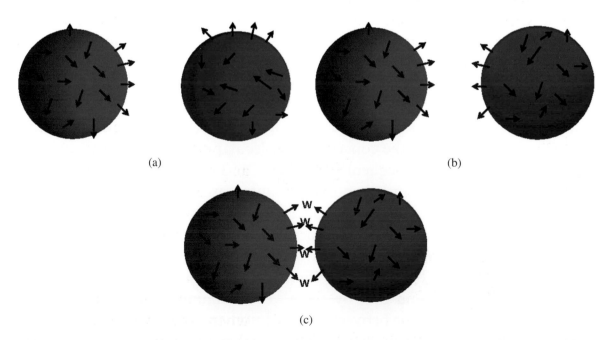

Fig. 64 Two proteins with small 'arms' (small arrow) (a) before rotation; (b) after rotation (c) bound by water bridges.

No one could even venture a guess. There was total silence in the class.

"Well, let me now rotate one protein. Can you now guess how these two proteins will bind?" asked the professor.

"Yes!" everyone shouted excitedly. It was clear now that after the rotation, at this particular orientation, the two proteins could be "glued" together by the water molecules.

"Now you can see that water not only provides the glue — in other words, the force that binds the two proteins — but also selects the precise surfaces of the two proteins that come together in the bound pair. In reality, the binding can be affected both by the shape — the lock and key — as well as by the water 'glue,' what I previously referred to as water stitching the two proteins together.

"I hope that you can now appreciate what we mean by molecular recognition. Let me proceed by giving you some examples where the action or the function of the protein crucially depends on such recognition.

"The simplest example is the binding of small molecules, referred to as a substrate, to the active site of the enzyme — similarly, the binding of a regulator to the regulating site on a regulatory enzyme. Without molecular recognition the enzymes would not act specifically on a specific substrate.

"There are many other cases where molecular recognition is vital for the function of proteins. Perhaps, the best example is the immune system. Here, scientists not only apply the concept of 'molecular recognition,' but the concept of 'memory' is also frequently used.

"Our bodies are constantly exposed to dangerous invaders. These could be toxic like snake venom, or insidious such as viruses or bacteria. The first line of defense is our skin, which is impenetrable to many substances. However, many insidious substances can penetrate our lungs or intestines. For these types of invaders, the body needs a whole army of soldiers that will identify the harmful invaders and destroy them. The immune system distinguishes between friend and foe, or between self and non-self. The whole idea of vaccination is based on our knowledge of how the natural immune system works.

"At the heart of the immune system is a molecule referred to as an antibody. This is a complex protein that can identify foreign molecules. A typical antibody has a Y shape, and the base of the Y has a constant structure while the two 'arms' of the Y are variable. It is this part that is responsible for identifying, binding and destroying any foreign invader. Once the body is exposed to a given bacteria, the immune system produces antibodies that are specific to that bacteria or any other *pathogen*, which is a disease-causing agent.

"The second time the body is exposed to the same bacteria, it is already prepared with the specific army or antibodies to fight back and ward off the disease. It is in this sense that we use the term 'memory,' as if the immune system 'remembers' a previously invading foe. Vaccination is essentially 'teaching' the immune system about a possible harmful invader. By exposing the immune system to a vaccine, a partially weakened form of the bacteria, the system identifies it and prepares an arsenal of antibodies to fight back any future attacks by the real, fully active bacteria.

"In some cases the immune system fails, unfortunately, and this can spell disaster for the whole defense system. These are referred to as autoimmune diseases. The HIV virus is an example. It was discovered that the HIV virus disrupted the immune system itself. As a result, it is very challenging for scientists to find a vaccine for this disease, given that vaccination is akin to 'teaching' the immune system. However, if the immune system fails then who is going to be taught? Therefore, alternative methods to fight back must be developed.

"I would like to mention one more important case of regulation where molecular recognition is important. This is the case for the 'expression' of specific genes in specific cells. First, let me say that a gene is a small segment in the DNA that can be translated to a specific protein. You remember that the DNA contains the information for all the proteins that are synthesized by the body. However, not all the proteins are needed in each cell. Some cells need one group of proteins; others might need another group. A cell does not expend time and effort on reading, transcribing and translating genes that contain information about proteins it does not need. Instead, it reads, transcribes and translates only those genes that produce the required proteins for that cell.

Fig. 65   Kofeau and his friends binding proteins.

"The job of selecting, activating or blocking genes is done by proteins called repressors. These proteins bind to a specific point near a gene that the cell does not want to read. The mechanism of binding and then blocking the expression of the gene depends on the ability of the repressor-protein to recognize the right genes it wants to block. For this reason, this system has been called a 'genetic switch,' that is, the repressor can determine which genes will be switched on or off in a given cell.

"Molecular recognition is also important in the construction of supramolecular assemblies of proteins. Let me mention only a few examples of such multi-subunit assemblies. Viruses are nothing more than a file of information in the form of the DNA or RNA wrapped in a package of proteins. One of the most studied plant viruses is the tobacco mosaic virus, TMV, which infects the leaves of plants. It's basically a single strand of RNA coated with over 2,000 identical protein subunits.

"I should note here that viruses are not considered living organisms mainly because their reproduction depends on their using the machinery of the host cell into which they penetrate. However, they do reproduce and mutate like any other cell, and in this sense, they have the characteristics of a living system.

Fig. 66   Tobacco mosaic virus (TMV).

"The main engine in which the information on the RNA is translated into protein is also an assembly of RNA and proteins. These are the ribosomes, which consist of RNA, or rRNA, and over 80 subunits of protein. It is in these ribosomes that the information carried out by the tRNA is translated, phrase by phrase into amino acids to produce the protein."

Professor took a deep breath before concluding.

"I have given you only a preview of the very important, ubiquitous phenomenon referred to as 'molecular recognition.' I hope I have given you enough examples to get you interested, and perhaps even research this exciting field one day. Our next class will be the last one for this semester. I will summarize what we have learned, and I will be glad to answer any questions you might have, if I possibly can."

15

The Extended Picture of the Central Dogma

A triumphant looking Professor Holmes entered the classroom on the last day of the semester. This time he did not have with him his usual teaching materials. As he always did, the professor had mixed feelings on the last day of class. On the one hand, he was happy and satisfied because he had fulfilled his mission as a professor. On the other, such occasions also made him sad, as he knew he would miss the lively interactions with his students, who he treated as if they were his own children.

"Today, we shall discuss some aspects of the overall view of the central dogma, with which this series of classes began," said the professor, showing the first slide. "The central dogma essentially consists of three elements: the replication of the information on the DNA, the transcription of the information from DNA to RNA, and the translation of information from RNA to proteins. As you can see, I used the term 'information' three times in those three steps. I will return to this and discuss in what sense the term should be understood, but for now I trust that you grasp the term 'information' in its intuitive sense, which is fine.

"In fact, there are many scientists who use the term 'information' in connection with what happens *after* the protein comes off the production line. For instance, you can find statements like, 'the information in the sequence of amino acids is further translated into a precise 3D structure of the protein,' and in a more general sense, 'DNA carries all the information for a specific living creature.'

"In some sense this is true, but we have to be careful about the meaning that we assign to this very loaded term. First, let me again point out that in each cell

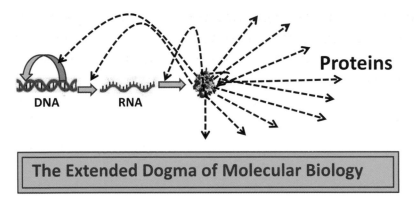

Fig. 67   The extended dogma.

of the body there is a copy of all the information about every cell. However, each cell *expresses* only a part of that information. More specifically, the DNA is made of units called genes. A gene is like one 'sentence' of the book, and it can produce one protein. In a specific cell, the whole 'book' of DNA is available, but only the chapters that are relevant to that particular cell are read, and the instructions written there are executed.

"Clearly, each cell is different. Blood cells are different from nerve cells, and these are different from liver cells, and so on. Therefore, each cell 'reads' only the part of the DNA that is relevant to it. We say that the relevant genes are *expressed* in each cell. This expression, or 'reading,' of the relevant information, and the 'ignoring' of the irrelevant information is also done by proteins, which are called regulatory proteins — like the so-called repressors that we discussed in previous classes.

"Thus, you can see that the very execution of the processes of replication, transcription and translation do not end when the proteins are produced. These proteins perform a multitude of functions. Some of them go back and execute and regulate the very processes that had produced them. They help the replication process, the transcription and the translation process.

"As we have seen, the machinery of protein production, the ribosomes, are also made almost entirely of proteins. This is a complicated and rather magical cycle. The DNA produces the proteins, and the proteins help produce the DNA, and they participate in the process of their own production. It's the classical question: Which came first, the chicken or the egg, the DNA or the protein? We do not know exactly the answer to this question, but perhaps there is *no* answer to this question!

"Some scientists speculate that in the beginning there were some primitive molecules, neither DNA nor protein, perhaps some kind of a primitive form of RNA, a molecule that could reproduce on the one hand, and at the same time had some catalytic effect on the replication process on the other hand. After millions of years, these primitive molecules evolved into more 'sophisticated' or more specialized molecules — some took the role of primitive DNA while others took the role of primitive proteins. After many more millions of years, both DNA and proteins diverged into very different molecules carrying different information and having different specialized functions. Don't forget that what we observe today is a result of many, many million years of evolution.

"Let me discuss the term 'information,' which I have used so many times in connection with DNA and proteins. Remember that the units that comprise the entire DNA are genes. Each gene produces one protein, and in this sense we can say that the information for that particular protein is *contained* in the particular gene. But we do not have to use the term 'information' at all. We can also say that the gene *encodes* the sequence of amino acids, without ever using the term 'information.' This statement is *a fortiori* true when we use the term 'information' in the context of the entire DNA and the entire animal. Here, one must be careful in using the term information with respect to the entire animal contained in its DNA.

"Think again about the instruction manual for producing a building. This hypothetical book not only contains instructions on making the floor, the walls and other parts of the house, but also contains information on how to construct 'window makers' and 'wall makers,' and so on, as well as how to construct a robot that will control all these processes.

We have implicitly assumed that the *conditions* for the production of all these things have been met — for instance, that there is a constant supply of raw materials, that the temperature is not too cold or too hot, and so on. Now consider the same book containing all the information to construct the whole building, but while construction is going on, there is a shortage of some raw material, say glass. What will happen? The 'machine' might continue to do its work, but that building will have no windows. This is a simple example. Other changes in the environment could be more detrimental. For instance, if water is in short supply then cement production is not possible; thus the process will come to a halt. On the other hand, if there is a flood, an over abundance of water, many of the robots will not be able to perform and the building, if started, will be half-done.

"You can think of hundreds or even thousands of factors that might result in some defective, deformed, half-done or unfinished building. The same is true for the information contained in the DNA and the process of constructing the entire animal. It is implicitly assumed that the external conditions of temperature, pressure and the supplies of raw materials are suitable for the process of building the entire animal. However, many things can go wrong, and many things do go wrong.

"Suppose that at some point of the process some raw materials are in short supply, or missing. In this case, the process might proceed but the end product might be a building with some defects. Thus, the entire information is there in the building's DNA, but the end product depends on the external conditions or the environment under which the whole process is undertaken. By the same token, under different conditions, the same DNA can lead to different dogs or cats, or even humans.

"If one substance is missing, you might get a dog with no hair, no eyes or no feet. Would one say that the same DNA contains information about eyeless, hairless or legless dogs? In more extreme cases, there might be no dog at all. So where did all the information go? As you can appreciate, this is an immensely complicated system. The same DNA of a specific animal can produce infinitely many *different* animals of the same species depending on environmental factors, such as temperature, pressure and concentrations of various external ingredients.

"Some conditions will produce a dog with defects, but nevertheless a functioning dog. Some conditions can lead to a dog of a different color, or with missing a toe. Others might lead to a dog that cannot survive, or perhaps no dog will be produced at all. Therefore, you should be careful when considering the meaning of the information encoded in the DNA. You must always remember that whatever the code is, it must be *read* and be *executed*, and this 'reading' and 'executing' depends on external conditions. Different external conditions — indeed an infinite number of different external conditions — can produce infinitely different products.

"So, DNA contains the information for the construction of the entire body. This statement is correct but there is no *code* that translates from each letter or phrase on the DNA to some feature in the animal. If such a code existed, it would be dependent on the environment, which includes temperature, pressure and the concentrations of countless molecules. This is equivalent to a code for each environment, which effectively means an infinite number of 'codes.'

"Let me summarize briefly what we have learned. The information contained in DNA flows from DNA to RNA, to proteins, to the final structure of the entire body. You can say the same thing without using the term 'information.' It is sufficient to say that a specific sequence of DNA is translated to a specific sequence of proteins, and the resulting proteins do everything else to produce the body, including the replication of DNA and the translation from DNA to protein.

"I hope you have grasped the ideas contained in the central dogma of molecular biology. You should realize that the discovery of each step in this dogma has taken the concerted effort of a huge number of scientists around the world. Yes there is always much more to be discovered, and I hope that this series of classes will encourage you to further study this interesting field. I wish you all a pleasant vacation. I'm looking forward to seeing you again next semester."

As Professor Holmes finished talking, the classroom spontaneously erupted in applause, and the students even got to their feet. Alice clapped louder than anyone. The impromptu standing ovation touched the professor deeply, and as left the classroom, he felt tears welling in his eyes.

16

Alice Bids Professor Holmes Adieu and Gets a Bonus from Kofeau

As the semester drew to a close, everyone in Alice's class had gone off to celebrate the beginning of the much-anticipated break. Alice, however, had one priority not shared by her classmates. She wanted to thank not only the professor but also Kofeau, who had done so much to enhance her knowledge. She had a special attachment to Kofeau, both the real one, who entertained them with his comical ways in class, and the virtual one, who had made her learning experience even more enjoyable and interesting.

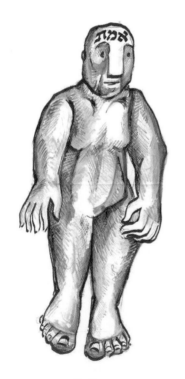

Golem

Fig. 68 The Golem.

Alice also desperately needed to satisfy her curiosity. Had the little monkey merely been mouthing what the professor had programmed him to say? Or did he indeed have a mind of his own? Were those his own views he had expressed? She remembered the story of the Golem, the strange being made of clay created by a rabbi in Prague that eventually became so uncontrollable that it turned on its master.

These thoughts and questions occupied Alice's mind as she made her way to the laboratory.

Also, although Alice was very keen to thank the professor, she also had an ulterior motive for her visit.

When she got there, she found the professor deeply engrossed in a mountain of papers.

"Good afternoon, Professor Holmes!" she said cheerfully. "I just came by to thank you for making this semester's lectures so interesting. I wanted to thank you especially for the simulated programs that you created, which were instrumental to my understanding of the difficult topics we dealt with."

Alice deliberately didn't mention Kofeau, knowing that he was simply part of the program designed by the professor.

"You are very welcome, Alice. I do appreciate your coming," the professor replied, "and particularly if the simulations were helpful."

He paused for a moment, just long enough for Alice to wonder what he would say next. It was quite an unexpected question.

"Would you also like to say goodbye to Kofeau?"

Alice was taken aback. She really wasn't sure whether the professor was being serious. Did he really want her to bid Kofeau farewell personally? Did the professor see the little monkey as someone making his own contribution and so deserving to be thanked separately?

"Why, yes of course!" Alice replied, trying to contain her excitement. "Kofeau was really very kind and helpful. I would love to pay him a visit — if he has some free time!"

"Well, you know where to find him," the professor said, grinning. "He's expecting you."

Alice's brain spun. How could he know that Kofeau was expecting her? The professor must have programmed Kofeau to expect her! This only went to show that Kofeau was just part of the professor's program, Alice thought to herself. But then what was the point in saying goodbye to Kofeau? Now she really needed some answers.

Lost in her thoughts, Alice drifted mechanically to the ExploCube and slipped on the goggles. Before she could think anymore about it, Kofeau's friendly face appeared.

"Hello Alice! How are you today? Do you have any questions for me?"

"No, thank you, Kofeau," Alice began. "I just came to say goodbye to…"
Without allowing Alice to finish, the little monkey interrupted her.

"If you don't have questions about the lectures, perhaps I can suggest an unrelated question, something that isn't about the lectures? I'd be very happy to suggest a topic to discuss, but I would be equally glad to answer your questions."

Alice was intrigued. She knew that what he had to say would be something interesting.

"What did you have in mind, Kofeau?" she asked.

"Well, it has something to do with the solubility of proteins," said the little monkey. "As you know, most of the proteins function in an aqueous media, so in that sense it is relevant to the lectures. But I gather that Prof. H didn't discuss the 'solubility of proteins' as a problem. Am I right?"

"No!" she replied firmly. "He never mentioned the problem of solubility of proteins."

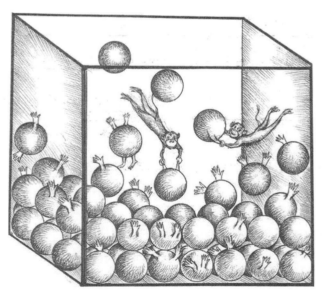

Fig. 69   Solubility of protein.

It was funny for Alice to hear Kofeau referring to "Prof. H" again. But she couldn't imagine why he would bring up a problem that was never discussed in the classes. Kofeau really did appear to have his own personality, Alice thought to herself, laughing silently at her own joke. Was Kofeau really thinking for himself, independently of his master? Was it just like the professor's story about the mysterious Golem?

"Well, you know that most proteins 'live' in aqueous media," Kofeau continued. "Have you ever asked yourself what makes the protein soluble in water?"

Alice considered the question for a moment. She recalled that some solutes had arms, and were therefore water loving and thus soluble; and some were water fearing, and therefore not soluble.

"Well, I never really thought much about the solubility of proteins, but as you have explained, and as you have demonstrated, the proteins are built up in such a way that most of the water-fearing groups are buried in the interior, and the surface of the protein has an abundance of water-loving groups, meaning that these groups extend many *arms* into the solvent. Therefore, the water molecules are happy to hold hands with these arms, and that is what makes the protein soluble. It is a lot like a micelle, where the hydrophobic groups are inside while the hydrophilic groups are outside."

Alice was proud of herself for coming up with a convincing answer to Kofeau's question.

"Excellent, Alice! I take my hat off to you!" exclaimed Kofeau. "Basically, that is correct. But let me tell you that what you have said is what most scientists would agree upon. Perhaps that is the reason why Prof. H did not pose that as a problem at all. However, I want to add one more factor that enhances the solubility of the protein which not too many scientists are aware of — perhaps not even Prof. H."

Half whispering his last words, Kofeau winked at Alice and they exchanged knowing smiles. It was their secret, just between the two of them. At that moment, Alice had completely forgotten about the program, about the professor's simulation, about the ExploCube. She had her own special connection with the little monkey — and it didn't have anything at all to do with Professor Holmes.

"Let me show you what I meant," said Kofeau, "when I said that it is true that the reason for the high solubility of proteins is the many arms extending from the surface of the protein into the water, but that is not the complete answer."

Kofeau pointed to a large nearby sphere that was floating in water molecules.

"This sphere represents a globular protein. As you can see, there are about 20 arms extending from the surface of the protein into the water, and each of these arms can hold an arm of a water molecule. As we have already discussed, this arm-holding with water molecules makes the protein more soluble. Here, the arms are far from each other so that at any given point in time an arm of the protein can engage in arm-holding, or establish a hydrogen bond, with one water molecule.

"Now, look closely what happens when I change the distance between some of the arms. You see that at a certain distance a *pair* of arms of the protein can be bridged by one water molecule. This should remind you of the role of water in protein folding and in protein–protein association. It turns out that each 'water bridge,' if you like — a single water molecule holding hands simultaneously with two arms on the protein — can cause an increase in the solubility by a factor of 100. You can imagine that in a real protein many arms can be bridged by water molecules. In fact, there is also a possibility that a water molecule can hold three arms of the protein, further increasing the solubility."

Kofeau moved the protein's arms so that they held onto a water molecule with three of its arms.

"To summarize, it is true, as you have said, that the arms contributed by water-loving amino acids of the proteins are responsible for the solubility of the proteins. But it is not only the *number* of arms that makes the protein so soluble. More important is the *distribution* of the distances between the arms that makes the 'water bridges' possible. This dramatically increases the solubility of the proteins, or if you like the extent to which the proteins are welcomed into the water milieu."

After his presentation, Kofeau waddled up to Alice, his long arms extended as if he was about to embrace her.

"Thank you, Kofeau," said Alice, adding softly, "It was very kind of you to explain the solubility of proteins to me. I was not aware that such a problem exists, as the professor never mentioned it…"

Alice realized she faced a dilemma. She had her secrets with Kofeau, but she felt she really ought to share her "discovery" with Professor Holmes. Fortunately, Kofeau seemed to have read her mind.

"By all means, Alice. Please feel free to discuss with Prof. H whatever you like," the little monkey said, smiling reassuringly.

And with that, though still a little puzzled, Alice bade Kofeau farewell. "I'll say goodbye for now then," she said. "It was a real pleasure to meet you."

Fig. 70　Alice bids Kofeau farewell.

In an instant, everything vanished before her eyes and Alice was back in the lab again. She peered out of the strange booth, and there was a relaxed-looking Professor Holmes, sipping his favorite tea concoction.

Alice got out from the booth and wondered what to do. Should she tell the professor about her latest experience with Kofeau? Or should she keep it to herself and honor her pact with the little monkey? The dilemma evaporated as soon as she heard Professor Holmes' first words.

"Welcome back, Alice," he said with a smile. "You must be wondering whether Kofeau has his own will or not."

"The truth is," replied Alice, relieved that the professor had saved her the embarrassment of bringing up the matter. "I have often wondered whether Kofeau was acting on his own, and I had come to the conclusion that you had programmed everything... But what was the reason behind making me feel as if... Why did you program Kofeau to act as if he had his own personality? Why did you make it appear that I was sworn to secrecy? Why did you want me to bid farewell to Kofeau at all?"

"Well, Alice," said the professor with a grin, "you were perfectly right to suspect that Kofeau had his own opinions — and sometimes even a different opinion from what was discussed in class. I did it purposely so that you would be more critical. Often one hears not one or two, but several opinions, some of them conflicting. I wanted you to not take everything at face value. I wanted you to be more critical and analytical. One should always try to determine which theory is more convincing. That is precisely the reason why, at times, I made Kofeau have different views from those I expressed in my lectures. I hope you can appreciate my objective in doing so. I'm sure that you have benefited from it."

"Yes, absolutely!" Alice answered, still digesting what the professor had said.

"Now, do you have any other questions?" said Professor Holmes. "I have to run a few errands for my wife."

"Oh, no thank you, professor," she replied. "I'm still a little overwhelmed. I must admit that at times I thought that Kofeau had turned into another Golem and you had lost control of him! But it has truly been a great learning experience for me. I'm very grateful for everything you have done — directly or through Kofeau. Thanks again, Professor Holmes."

Fig. 71   Alice with Professor Holmes.

Alice hesitated for a few seconds, realizing the absurdity of what she was going to say.

"Programmed or not, Kofeau's performance was superb! Please extend my greetings to him. I hope to see more of Kofeau in class next semester."

Alice looked at her watch as she was leaving, only then noticing how late it was. She had less than half an hour before she was supposed to be meeting her mother at the mall across town. As she started running, she suddenly realized how much she was looking forward to spending a few weeks with her mom, and how much she needed to relax after all the information she had absorbed over the semester. She would need time to sort out in her mind the most important things she had learned — from both of her teachers.